9/11

Seattle and the Roots of
Urban Sustainability

History of the Urban Environment

Martin V. Melosi and Joel A. Tarr, Editors

Seattle and the Roots of Urban Sustainability

INVENTING ECOTOPIA

Jeffrey Craig Sanders

University of Pittsburgh Press

Published by the University of Pittsburgh Press, Pittsburgh, Pa., 15260

Copyright © 2010, University of Pittsburgh Press

Manufactured in the United States of America

Printed on acid-free paper

10 9 8 7 6 5 4 3 2 1

Library of Congress Cataloging-in-Publication Data

Sanders, Jeffrey C.

Seattle and the roots of urban sustainability : inventing ecotopia /

Jeffrey Craig Sanders.

p. cm. — (History of the urban environment)

Includes bibliographical references and index.

ISBN-13: 978-0-8229-4395-2 (cloth : alk. paper)

ISBN-10: 0-8229-4395-6 (cloth : alk. paper)

1. Seattle (Wash.)—Environmental conditions.

2. Environmentalism—Washington (State)—Seattle—History—20th century.

3. Urban ecology (Sociology)—Washington (State)—Seattle—History—

20th century.   4. Sustainable development—Washington (State)—Seattle—

History—20th century.   5. Seattle (Wash.)—Social conditions—20th century.

6. Neighborhoods—Washington (State)—Seattle—History—20th century.

7. Suburbs—Washington (State)—Seattle—History—20th century.

8. Suburban life—Washington (State)—Seattle—History—20th century.

9. City and town life—Washington (State)—Seattle—History—20th century.

I. Title.

GF504.W2S26 2010

307.7609797772—dc22   2010011363

For Kate

Contents

Illustrations

Acknowledgments

The academic life can defeat our best efforts to dwell. Over the last ten years I have made many places home. This book is shaped by that experience and by the people who have helped me to think more deeply about place.

In New Mexico I found a group of scholars who helped me the most. David Farber taught discipline and freedom. I have benefited from his intense and careful mind and his open intellectual curiosity. He asked the critical questions at the right times and has been a generous and constant supporter of my work and career. I owe Virginia Scharff deep thanks. Her commitment to students and colleagues, her prodding and humor, and her dedication to environmental and American West history taught me invaluable lessons about community. With Sam Truett I got to know the field of environmental history in the best way possible, including perambulatory discussions at Armendaris Ranch in southern New Mexico. I'll never forget Sam's generosity. Chris Wilson helped me think more carefully about both visual evidence and everyday landscapes. Thanks also to Linda Hall, Vera Norwood, Pat Risso, Ferenc Szasz, and especially Timothy Moy. Tim modeled a down-to-earth approach to excellent teaching and critical thinking. We miss him.

Many big and little institutions provided temporary shelter. For several years I also found a home working with an amazing group of people under the guidance of a talented editor and professor, Durwood Ball, at the *New Mexico Historical Review*. Thanks to those who helped to shape my foundation. James Martin deserves special thanks for thinking with me over the years and for sharing many a sustaining meal of *carne adovada* burritos at the Frontier Restaurant; Eric Loomis read drafts and took walks; John Herron suffered many "very Sanders days," and I was lucky to have him as a critic and sounding board. Thanks also to Jonathan Ablard, Javier Marion, Ev Schlatter, Lincoln Bramwell, Judy Morley, Jennifer Norden, Sarah Payne, Amy Scott, Kim Suina, Blair Woodard, Tiffany Thomas, Kyle Van Horn, Catherine Kleiner, Jeff Roche, and Rob Williams—I feel lucky to have shared time and ideas with such a strong group of scholars and storytellers.

Numerous librarians and archivists at public institutions provided temporary quarters. My deepest thanks go to Anne Frantilla at the Seattle Municipal Archives, who humored me and helped me find solid footing in the records collections at the old and the new Seattle Municipal Archives. Thanks also to all the staff and librarians at the University of Washington branch libraries and particularly the University of Washington Special Collections. Trevor Bond and Cheryl Gunselman at Washington State University Manuscript, Archives, and Special Collections have been wonderful new colleagues. They, along with Mark O'English and Pat Mueller, were generous with their time reacquainting me with the Washington Tilth papers. The Puget Consumer Cooperative allowed access to their collection of newsletters from the past forty years. The Museum of History and Industry, a Seattle treasure, supplied much of the visual evidence that was important to this project. In Maine I found safe harbor for two years. Thanks especially to Elizabeth DeWolfe, whose professionalism is unparalleled and who perfectly models the balance of scholar and teacher. Thanks for taking a chance on me.

A broader community of far-flung scholars read the manuscript at different stages and offered thorough and generous critiques. Thanks especially to Carl Abbott for his kindness and close readings; to anonymous peer reviewers at the *Pacific Historical Review;* to Raymond Mohl, Martin Melosi, and Cynthia Miller; and to the peer reviewers for University of Pittsburgh Press. Thanks to the *Pacific Historical Review* for allowing me to reprint a chapter that was originally published as Jeffrey C. Sanders, "The Battle for Fort Lawton: Competing Environmental Claims in Postwar Seattle," *Pacific Historical Review* 77, no. 2 (May 2008): 203–36. Matthew Klingle shared his considerable understanding of Seattle's environmental and political history and asked incisive questions early and later in the process. I wish I had even half of Matt's integrity, energy, and skill. He is a wonderful model. Mark Feige made insightful comments late in the process. This work also benefited from a unique conference at the German Historical Institute called "Nature in the City." Thanks also to those who told me their stories and shared their sense of place and history: Jody Aliesan, "Uncle" Bob Santos, Mark Musick, Elaine Ko, and Carl Woestwind.

Now settled on the Palouse, I've been welcomed by a strong community of scholars and new friends. I thank Rob McCoy, who read chapters, and Jesse Spohnholz, who offered camaraderie; Ray Sun, for his support of junior faculty; the College of Liberal Arts, for the Meyer Project Grant that helped to pay for important permissions and follow-up research trips; and Pat Thorsten-Mickelson, for making it all run smoothly. Adam Sowards has been a careful reader and thinker over on the other side of the border, at University of Idaho.

Finally, thanks to my families: my Arizona family, Charles and Lynn Ffolliott, for their support and good humor, and my Maine family, Peggy Forster, who may not be a rock but has been a lifeline and has taught me a great deal about place. I've had the unquestioning support of my Seattle family. D'Arcy Hyde and Kristin Berkvist were there for my city fix and for new eyes on my old city. Thanks to Kristin Hyde for years of good conversation about food. Kenan Block, another "native" like me, has been my link to the Seattle past and present. My first intellectual community, Barbara and Craig Sanders, as well as my sister Karen's family, Doug Balcom, Zan, and Claire—all have supported me in innumerable ways materially, emotionally, and intellectually. My parents provided the raw materials for an intellectual life with art, food, argument, and tolerance.

Kate Ffolliott and I have made homes together coast to coast with our intrepid daughters, Sarah and Ellie. Kate has my deepest gratitude for just listening to me and for, as Toni Morrison said in *Beloved*, "putting her story next to mine."

Seattle and the Roots of
Urban Sustainability

Prologue

"The Battle in Seattle"

On November 30, 1999, Seattle's wet downtown streets echoed with chanting voices and festive horns. That fall, Seattleites looked up from their lattes to witness a carnival of protesters and impassioned street theater descending on their city. As Christmas drew near and shoppers filled the streets, dignitaries of the World Trade Organization's (WTO) ministerial conference converged on Seattle. The WTO conference drew together trade representatives and nongovernmental organizations (NGOs) from wealthy member-nations around the world to plan and negotiate far-reaching trade agreements affecting economies, large and small, across the globe. The proud metropolis on Elliott Bay had beat out forty other cities for the honor of hosting an event that would convene over five thousand delegates from 134 countries for the largest trade summit ever held in the United States.[1]

The host city, Seattle, offered a friendly venue that local leaders believed could contain both the expected protests and the business of world trade and its dignitaries, including a visit by President Bill Clinton. Seattle's mayor, Paul Schell, a former community development official instrumental in Seattle's rise from a provincial West Coast backwater in the 1960s to a "world-class," high-tech trade capital in the 1990s, captured the mood of anticipation. "Seattle likes hosting these kinds of things," he told reporters. "We see ourselves as an open city, a center of creative debate." Rather than close off Seattle's grid of hilly streets to forestall trouble, the mayor approved large demonstrations and even an area downtown near the main WTO venue, the Washington State Convention Center, as a meeting place for protesters. One of the many activists who came to Seattle then remarked how open the city seemed; "It's almost like giving us a home field advantage," he said, noting the city's "great labor energy" and "all these environmentalists."[2]

The first giddy moments of direct action politics that day culminated in a long-planned march through the heart of the business district among the gleaming skyscrapers and recently revitalized upscale shopping areas. Crowds filled the streets, not far from the shores of the busy international shipping hub on Elliott Bay. A collaboration between the AFL-CIO and environmental groups in particular, the procession filled the streets with over forty thousand people in the day's main event.[3] Marching in unity, trade unionists protested unfair global labor laws and sweatshops in solidarity with environmentalists dressed as sea turtles and stilt-walkers in monarch butterfly costumes decrying the rollback of environmental laws protecting endangered species.[4] The unprecedented public alliance of such disparate groups impressed observers and defined the hopeful and unified tone of the first protest. The *New York Times* described the mixture: "The protest groups ranged from the well-known, like Friends of the Earth and the Humane Society, to the obscure, like the Ruckus Society and the Raging Grannies."[5] The direct action protesters were indeed a diverse and motley group, challenging aspects of globalization from multiple political positions.[6] Journalists reveled in what they saw as an unlikely alliance of tree-hugging Sierra Club members and blue-collar United Steelworkers who, by the end of the day, would perform a "Seattle Tea Party" by throwing imported Chinese steel and hormone-treated beef, "goods they viewed as tainted by the trade body's decisions," into Puget Sound, just blocks from tourists shopping at the Pike Place Market. Echoing another famous anticolonial economic resistance movement, the crowd shouted, "No globalization without representation" as they heaved away.[7]

The quickening pace and geographical disparities of globalization were appearing increasingly stark to many activists by 1999, and the protesters in the streets of Seattle suggested new connections, the contours and potential power of a contemporary environmental movement that would resist it. In their diversity, these critics of globalization manifested the "three Es" of the post-1990s sustainability movement: the overlapping concerns for environment, economy, and equity. The United Nations World Commission on Environment and Development articulated this conception of environmentalism officially in 1987; it was later reaffirmed by the so-called Rio Conference (or Earth Summit) in 1992.[8] Each body's official language stressed the idea of "sustainable development." While journalists may have been confused about the strange bedfellows marching together in Seattle, these protesters perfectly reflected the potential of contemporary environmentalism with their concerns for environmental justice on a global scale, what the UN's Rio Conference principles described as the link between "sustainable development" and the "essential task of eradicating poverty." On the streets of Seattle, advocates of democratic reform, social justice and human rights, environmentalism, and labor came together in one massive, if temporary, coalition during the protest. Together they revealed to the public, and perhaps concretely to themselves for the first time, productive affinities among far-flung places and diverse activists. They highlighted connections among environmental and human problems that once may have appeared separate. These activists embodied the vision of sustainability that pronouncements in official documents only suggested. A decade's worth of UN meetings had helped to produce an official language, but the protesters on Seattle's streets brought legs and the energy of grassroots organizing to these declarations. The street scenes in Seattle therefore captured an explosive, promising moment in global environmental politics, one that had been building for some time. These events became a renewed rallying point for the global Left as well, a fresh model of coalition building. As members of the World Trade Organization met to articulate *their* global vision, this other, grassroots coalition seemed to gain momentum in the streets of Seattle.[9]

Later that day, though, things turned a bit ugly. Despite the city's hospitable stance, or perhaps because of it, both the trade conference and the peaceful protests were soon eclipsed by a chaotic scene. The city had become, in the words of the *New York Times*, "a giant protest magnet."[10] The growing throng engulfed the streets in what would come to be called the "Battle in Seattle."[11] In a more confrontational mood, some in the mélange of protesters blocked intersections, linking their arms inside metal tubes to thwart movement on city streets. These peaceful, yet unbending protesters—some dressed as squirrels or endangered animals—successfully halted traffic at key intersections, blocking dignitaries' access to the conference site. The direct action aroused an intense response from Seattle police, who bruised protesters and bystanders with volleys of rubber bullets and smothered even peaceful protesters and bystanders in a cloud of tear gas.[12] But the most memorable moments, or at least the ones captured and replayed a million times around the world, featured masked, black-clad anarchists creating chaos without restraint. As Seattle police rounded up peaceful protesters in one part of the shopping district, they neglected to notice a group of anarchists who broke from the larger crowd yelling, "Anarchy!" and "Property is theft!" The anarchists were no less organized than the other activists who had trained for weeks leading up to the conference. Equipped with walkie-talkies and a plan of attack, the small group seized the moment of police distraction. As news crews and international journalists looked on, the men and women in black overturned dumpsters and lit fires, to the horror of early Christmas shoppers. They lifted metal newspaper boxes off sidewalks, sending them through the display windows of Seattle's upscale downtown shopping district. The crowd thrilled, but mostly chilled, as protesters climbed to the top of the Nike Store entrance and pried the N from the store's sign. In an effort to police themselves, other activists tried to shout the anarchists down, and some veteran activists found themselves in the unlikely position of protecting Nike, McDonald's, and the Gap, the very targets in their political messages about globalization's dark side expressed earlier that day. To their chagrin, this giddy, destructive, iconic image of Starbucks' windows shattered into pieces on the sidewalk stuck in the public's collective head.[13] The targets, Starbucks and Nike, were both "local" companies—that is, based in the Northwest—but both enjoyed an imperial reach that made them perfect symbols of "unsustainable" practices in 1999. The anarchists underlined, with little irony or subtlety, the complicated stakes and tangled threads of global consumption.

Many Seattleites could not believe their eyes. The *New York Times* observed, "as tear gas wafted near the city's venerable Pike Place Market, a simple question was on the lips of many people, especially angry merchants: Why, exactly, did we decide to be the host of these global trade talks?" A forty-year-old computer technician and native of Seattle told one journalist, "I mean, here we are in what is supposed to be the happiest, mellowest city on the planet, and look what's happening." Mayor Paul Schell, the target of prodigious criticism after the events, expressed similar sentiments. Abandoning his initial hands-off attitude, the mayor felt compelled to call in the National Guard, lock down forty blocks of the city center, and order a seven o'clock curfew, much to his disappointment. He noted that his administration included "people who marched in the 1960's," adding, "The last thing I wanted was to be mayor of a city that called in the National Guard."[14] Schell was certainly not the only one to be thinking of the 1960s when looking at the city in the aftermath of the events—good feelings punctuated by violence. The comparisons were unavoidable for many journalists as well, who could not help but invoke parallels to the street theater and dashed hopes of that decade. The protest was "like the 1968 Democratic Party convention in Chicago," reporters repeated.[15]

Yet what history or explanation is useful for understanding both the intensity of these events and, more important, the diverse constituencies that found common cause on the streets of Seattle? One local narrative, the most self-serving for Seattleites at the time, was simply that the city had been invaded by outside agitators. Bad people came to a good town. Seattle, like other world-class cities, was merely the hapless target of an international protest movement. After all, many of the anarchists were from Eugene or Portland, Oregon, havens for young radicals, rabid animal rights groups, and monkeywrenchers such as the Earth Liberation Front—people who might resent their neighbor to the north.[16] But this was too easy, a line of reasoning that ignored the complicity of comfortable, "progressive" cities in distant degradations.

Longtime veterans of urban politics such as Schell must have wondered how Seattle's image as an innovative, democratic, environmentally conscious, progressive city could have been tarnished so easily. How did his city become the Chicago '68 of collective millennial memory? Certainly, Seattle's smugness made it a "magnet" for protesters in the fall of 1999. But a deeper sense of the past provides important foreshadowing of these events. More than simple comparisons to the clichés of the 1960s-era "revolutions" and violent protests are necessary to make sense of the pent-up energy unleashed and targeted at this otherwise "mellow" city.[17]

The explosion *in* Paul Schell's city, after all, resulted directly from historical changes that began happening *to* his and all U.S. cities soon after World War II. In particular, when WTO protesters spoke to themes of uneven development and unequal environmental consequences on a global

5

scale, they appeared to be offering a new vision of environmental politics. Yet the connections they drew and the holism their organizing embodied owed a great deal to an earlier era of profound spatial disruptions and uneven development in the postwar American metropolis.[18]

Grassroots actors in the 1960s and early 1970s challenged freeways and urban renewal; fought to improve neighborhoods marred by poverty and racial segregation; contested the meaning of open space; and sought food security by building new, alternative networks of local production and consumption. In doing all this, they laid the crucial groundwork for this later global politics. This earlier period held both the spirit of possibility and deep contradictions that can explain the stunned reaction to the environmental politics on display during the "Battle in Seattle." In particular, the inherent tensions that propelled the urban environmental movement beginning in the early 1960s helps to explain the movement by 1999.

In the following chapters I argue that between the 1960s and the 1980s, cities provided an important crucible for contemporary environmentalism, explaining the emergence of the ubiquitous—though still contested—idea of sustainability and a movement that now finds a global audience. Journalists and other observers in 1999 were surprised to note these links among diverse political constituencies and interests, but this more complex picture and its set of actors have nevertheless been at the core of the movement from the beginning, driving environmental politics and grassroots activism since the early 1960s and before.[19]

The Battle in Seattle reintroduced important and long-standing questions—for the public and environmentalists alike—about the locus of political power, especially the different scales by which political power is measured and methods by which it is achieved and resisted. Urban environmental activists began asking these questions and testing possible answers amid the so-called urban crisis of the mid-1960s. Far from being an aberration in "a city with a tradition of liberal politics and civility in discourse," one that had, according to journalists, "a decidedly favorable bent toward environmental causes," the events in Seattle made sense in the light of enduring contradictions in American political culture and society that first erupted during the 1960s. These politics manifested in people's entwined struggles over specific environments in U.S. cities.

Eco-friendly Seattleites were surprised to find their city—their perfected blend of nature and urbanism—under attack and faced hard questions about the way they had reached such perfection and the costs incurred. Seattleites have long congratulated themselves for their city's celebrated relationship to its natural setting. Nearby mountains, salt and fresh water, and salmon shape Seattle's image as home to hikers and other outdoor enthusiasts. Urban and environmental historians have helped poke holes in this kind of civic boosterism, showing how people and the environments they build in places such as Chicago, Seattle, or Los Angeles have always made difficult compromises with the natural world. The WTO protests certainly bumped hard against Seattle's constructed self-image, forcing open-hearted Seattleites to question how this "mellow," nature-loving place—a prosperous blend of leisure, consumption, and environmental awareness—could have become the target of such rage, criticism, and violent reaction. Yet it is not a narrative of failure but one of compromises. If Seattleites were forced to face their contradictions in 1999, they were facing the shared contradictions of a postwar environmental politics that by the end of the millennium had forgotten its scrappier origins. This study begins to pull to the surface this partially submerged history of popular environmentalism, which derived strength from the urban social, political, and cultural contests in the era of postwar metropolitanization.[20]

On the streets of a host city that was on top of the world at the end of the millennium, WTO protesters returned to hard questions about building a sustainable planet: they underlined the drastic contradictions and disparities of uneven economic development; they linked social justice and racism to their environmentalism; they championed democratic, local, and grassroots challenges to concentrated power; and they highlighted the links between consumption and environmental politics. Yet these ideas are not new; they existed long before the WTO. In fact, as urbanites struggled with the shifting landscape of postwar metropolitan restructuring during the 1960s, these same ideas were always central to that era's environmental movement, though mostly ignored until recently. Metropolitan areas hold the clues to this history and the material grounding of these ideas.

The journalists who observed such a diverse array of activists working together in end-of-the-century Seattle were amazed in part because historians and popular treatments of the era have often tended to narrate the story of environmentalism as a mainly male, white, middle-class movement, or one focused on the management of wilderness and distant habitat, frequently couching their narratives in abstract terms. Often this movement's

7

story is divorced from the day-to-day experiences and politics of the cities, where most people have lived and worked and learned to care intensely about their nearby environments. The subsequent chapters focus attention on this local and lived experience as the starting point for understanding how people began to define their environmental politics and concerns in an urban setting. In addition, few histories have linked environmentalism to broader movements for social change, especially the grounding of environmentalism in the material and social conditions after 1960. Fewer still show how this movement and its ethos resulted from broader currents of national events during the second half of the twentieth century. Only recently have environmental historians begun to explore the role of people of color in this movement, for instance, or how this movement drew energy and expression in conjunction with the struggle for civil rights and gender equality after the 1960s.[21] Seattleites' attempts to grapple with a changing urban environment after 1960 illuminate the diverse origins of environmental activism and the multiple political and social contexts that gave it shape in American cities. Though the WTO protests could have happened anywhere, and indeed similar explosions of protest follow the organization wherever it alights, the events in Seattle, especially the diverse constituencies and holistic (even global) expression, suggest a history yet to be told.

Recovering this urban political genealogy of sustainability requires returning to the same city streets that protesters blocked. Like much of American politics and social change, the story plays out at a more micro level, affirming the notion that all politics, or social change, happens locally, a tenet of contemporary sustainability activists as well. Back in the 1960s, the area near Pike and Pine Streets where WTO protesters would later break windows had no Starbucks or Niketown dominating the landscape, just plenty of small, cheap cafés and restaurants. If any local businesses had achieved a global reach into sweatshops, rainforests, and coffee plantations in the developing world, such connections had yet to be acknowledged, much less bemoaned by critics of globalization. When the first Starbucks appeared in the Pike Place Market, with its burlap sacks of beans from exotic locales, it suggested a very different set of associations and sense of global citizenship than it would thirty years later. Yet with those same sacks of beans, the "whole foods" and do-it-yourself ethos at the city's public market, people would begin a process by which they would link their consumption to their environmental politics.

Just as fin-de-siècle protesters challenged the emerging spatial and economic order of global capitalism, activists in American cities during the early 1960s confronted a similar order on a national and local scale. On the streets where the anarchists attacked tony shops and multinational chain boutiques, there had once stood blocks of residential hotels, or single resident occupancy (SROs) hotels.[22] Lots of people, many of them poor, could still afford to live in the more densely populated downtown after World War II and in the first-ring suburbs of Seattle in the early part of the decade. As did other central cities during the 1960s and 1970s, however, the area quickly transformed as part of postwar economic restructuring. Following a pattern common throughout the United States, local business and government alliances in such cities, armed with federal subsidies, encouraged freeway construction, introduced urban renewal schemes, and systematically demolished low-cost housing while stimulating and quickening the pace of suburban growth. The historian Lizabeth Cohen has dubbed this set of swift and traumatic changes to the American landscape the birth of a "consumer's republic"; the resulting society catered to the automobile and middle-class whites and emphasized new forms of mass consumption and new definitions of citizenship. This crucial formative context—as profound in its way as the contemporary disruptions of modern global capitalism—helped to shape people's attitudes toward both the urban and the exurban environments. In Seattle, the contested landscape of this consumer's republic—who was included or excluded, who had the power to shape it, and what it would look like—formed the center of civic debates over the future of the American city.

To illuminate these important precursors to the contemporary sustainability movement, *Seattle and the Roots of Urban Sustainability* recasts and expands terrain that has often constituted the subject of environmental activism and concern. Americans began to worry about both natural and cultural resources at the same time. During the 1960s and the leaner times of the 1970s, Seattleites became increasingly agitated, politically active, and creative in their responses to the era's shifting material and cultural ground. The evolution of this urban-based environmental consciousness in Seattle and the energy of the postwar environmental movement in the city's neighborhoods became a true urban phenomenon.[23]

The postwar metropolitan context offers a sharper image of the critical links between urban history and popular ecology. The travails and experimental energy generated in the city between the 1960s and 1980s birthed innovative solutions to both urban and environmental problems. Urbanites across the spectra of class and race became concerned about nature at the same time that they became concerned about the future and meaning of the urban form itself. This environmental activism and thought could not be separated from the local and national movements for civil rights or gender equality in the changing city. Like Chicago at the turn of the twentieth century, Seattle during this time became a laboratory where diverse actors linked social change to environmental change; where they tested the emerging ideas of sustainability in the immediate spatial terms of their neighborhoods, open spaces, and public gardens; and where they ultimately reworked the meaning of home.[24]

Public contests over the idea, function, and physical form of the city evolved as the process by which Seattleites began to negotiate the earliest contours of what would become a vision for the sustainable city. After World War II, activists and downtown planners clashed over the proposed demolition of the Pike Place Market, an area that several groups identified as "the heart of the city" during the early 1960s. Changing perceptions of the city, perceptions shaped by the lived experience of place, federal leadership, and dollars, as well as an emerging class of urban experts, laid a crucial foundation for an explosion of neighborhood-level activism and planning. Beginning with proposed urban renewal at the Pike Place Market, different constituencies in Seattle tried to control, remake, and defend their ideal visions of the urban environment in response to postwar economic restructuring. Consumption and consumers sat at the center of these debates as Seattleites argued about the meaning and function of the city. Other cities struggled with similar changes at the time, but Seattle's earliest efforts to define a livable and democratic city clearly displayed a nascent urban environmental movement and the roots of sustainability.[25]

The Pike Place Market controversy, then, introduces key elements of an emerging urban environmental ethos in the context of postwar consumer culture. A simultaneous and less familiar example of urban environmental activism occurred in Seattle's Central District; rather than the contested "heart of the city" at the market, mere miles from downtown, this conflict occurred at the periphery of power during the War on Poverty. Just as WTO protesters asked incisive questions about the different scales of political power and global environmental problems, everyday citizens in Seattle's poorest and most racially segregated neighborhoods had been trying to understand and account for such connections and disparities on the local and national scales for some time. This ongoing grassroots effort to define a more equitable and healthy environment after 1960 serves as a case study for the movement's crucial and overlooked urban context.[26]

Shifting the focus to the roiling energy of neighborhood-level activism expands the set of players engaged in defining the agenda of the sustainable city. In fact, these places and less familiar players gave the movement shape and power, even as the mainstream, post–Earth Day movement came to be associated with white, middle-class actors and privilege by the late 1960s. Tracing this history therefore tells a different story of modern American environmentalism, its key actors, its origins, and its trajectory. Seattle was the first to participate in the citizen-organizing efforts of Model Cities, a late War on Poverty program, which mandated citizen participation in the planning process, and its participation helped to spur an array of decentralized, locally defined grassroots environmental activities in the Central District area. Building on the strong and diverse organizational structures of the civil rights movement in Seattle's Central District, the neighborhood's citizen-defined environmental planning put into practice many of the holistic ideas that would later characterize the sustainability movement while charting a course that environmental justice activists, advocates of New Urbanism, and city officials would follow in later decades. Model Cities organizing continued to influence the culture of the citizen-planning process in Seattle after 1970, especially as the program expanded to other parts of the city. In this way, the Model Cities program forms a crucial link between the racial and economic justice concerns of the civil rights era and the different forms of organizing that followed during the era of popular ecology. In their work to level the playing field of the city—economically, socially, and environmentally—activists in Seattle's neighborhoods prefigured the concerns of the movement on a global scale.

White, middle-class actors often dominated the production of space and the scope of proper environmental concerns during the late 1960s, but these more mainstream efforts to preserve nature or create urban parks occurred in a broader social and political context. At first glance, the preservation of the Pike Place Market and the neighborhood politics of the Central District are not immediately recognizable environmental issues.

Similarly, what appeared to be straightforward preservation efforts to create and define the uses of "open space" or habitat in Seattle were as much about nature as they were about anxiety regarding urban social changes and conflict. Pike Place Market defenders spoke of ecology and democracy, while nature's defenders spoke of poverty and social control. Different groups of Seattleites struggled for the power to shape and to control the city's green spaces in the midst of its transformation to a postmodern city, from a producer- to a consumer-oriented city conducive to specific forms of leisure and consumer activity.

These contested ideas of environmentalism were sharply focused in a locally famous late-1960s contest over public space at Fort Lawton, a decommissioned army installation in an upscale neighborhood. In part, the emblematic conflict pitted urban Native American activists against white, middle-class women in the Audubon Society, as well as wilderness enthusiasts. Oppositional actors and movements—including the civil rights, anti–Vietnam War, and Red Power movements—added a variety of voices, methods, and organizing tools that inspired vastly different kinds of activists to define the meaning of their environmental activism in a changing city by decade's end. Ideas of nature and of city were inextricably intertwined, and at Fort Lawton the contested nature of sustainability—the overlapping concerns for equity, environment, and economy—was most obvious. Native Americans at the fort used ecological rhetoric to define their needs for social and cultural space in the city. Similarly, white, middle-class bird advocates claimed new spaces in the city to foster wilderness and wildlife but also to express their own authority and desire for social control. And all activists claimed their authority to produce space in ways that revealed racial and class anxieties and privileges. Competing plans for the new open or green space, then, revealed the diverse and sometimes opposed political energies unleashed by the civil rights, antiwar, and women's movements.[27] Building on the market debate and Model Cities organizing, these political movements had lasting consequences for the way activists understood a healthy city. Environmentalism—as an attempt to control, order, or shape an ideal of urban or natural space—could not be separated from the immediate political struggles of the time. These different environmentalisms in Seattle, like the WTO protests, did not burst on the scene on Earth Day 1970. Rather, this movement took root in the productive soils of postwar social change.[28] The WTO showdown evoked this legacy and reflected many of the concerns, rhetorical modes, and insights of this early precursor to the debate about the definition of sustainability.

Both radical anarchists and more staid critics of globalization during the WTO protests targeted local businesses with global reach. Like the Native Americans who occupied Fort Lawton, these critics of globalization on the streets of Seattle hoped to remind the world's primary consumers of their relationships to the social and environmental health of people and places beyond their cities' limits as well as within. But again, the insights, language, and tools of this global movement were tested and practiced by urban activists of an earlier period.

The 1960s and 1970s counterculture played a crucial and understudied role in transforming public expectations for Seattle and for food consumption in the region defined as ecotopia. Co-op members, community gardeners, advocates of food security, appropriate technology advocates, and others created a lasting alternative network of production and consumption in the region. In particular, these activists, employing ideas of "steady state ecology," had already mapped out their own "bioregional" food system of small-scale and alternative production and consumption in the Pacific Northwest. Here again, activists revealed the critical role of consumption in shaping the politics of early urban environmentalism, the precursor to the contemporary local food movement. Taking a position contrary to the popular notion that people who identified as part of the counterculture simply dropped out or left the cities behind after the late 1960s, I expand on recent scholarship to show how counterculture environmentalists played a crucial and enduring role in shaping the tools and ideas of the contemporary movement for sustainability and environmentalism beginning in the 1960s. Because counterculture interventions and practices ranged widely—including how to buy, how to consume, and how to produce alternatives to mainstream culture—this study highlights the counterculture's distinct contribution. Yet these activists drew directly from the fertile ground that civil rights activists and other urban political actors cultivated. Benefiting from and expanding on the decentralized and creative approaches pioneered in the Model Cities neighborhoods and aided by a burgeoning counterculture movement on the West Coast, these actors began to imagine an environmental politics that explicitly linked urban homes and neighborhoods with a broader ecological system as part of their regional activism.[29]

Of all the groups attempting to control and redefine space in the city, then, counterculture environmentalists played perhaps the most significant roles by popularizing a critique of the status-quo city based on their application of "ecological" principles. Beginning in the early 1970s, the work of the Northwest Tilth Association, a network of "environmental heretics," food activists, and counterculturalists in the Pacific Northwest, laid the critical groundwork for an organic food economy in the region and helped to disseminate ideas about bioregionalism and do-it-yourself, decentralized, and low-energy agriculture inside and outside the city. Tilth's alternative production and consumption highlights the important thread of postwar consumption in urban environmental activism and illustrates how activists contended with the politics of consumption in forming new environmental attitudes. By the early 1970s, alternative modes of production and consumption in the "bioregion" sat at the center of an environmental critique of the city's remaking as activists sought to create a more just and equitable environment.

The ways in which activists remade the physical and cultural meaning of home and community in the specific urban context of the 1970s demonstrates the fine line between public and private acts of environmentalism. The gentrification process in Seattle's Fremont district and the changing idea of the urban house show the shifting urban policy atmosphere of the 1970s as the War on Poverty gave way to Nixon's "New Federalism," an era when drastic cuts in urban funding encouraged private solutions to problems. Many Seattleites made their private spaces the center of environmental politics in this context, an effort to create a "counterlandscape" of energy, food, technology, and society.[30] Numerous activists hoped they could challenge the dominant urban and rural relationship and make each of these places more "sustainable."[31] In Seattle, people began to think about the broader environment in relation to what they bought and sold, as well as what they grew and what they recycled, composted, or threw away. In the process, the green counterculture contributed to the reshaping of urban space, urban practices, and people's lasting expectations about what the contemporary city might look like. During this period, the garden, home, and neighborhood became powerful sites for the reconception of urban life. At the same time, these changes revealed the shortcomings of an urban environmentalism that increasingly sought private solutions, echoing a broader retreat from public life by the early 1980s.[32]

Seattle's multiple and decentralized experiments with the urban form became a persuasive package of urban ideas that powerful decision makers and professionals in the city hoped to adopt and even institutionalize by the late 1980s and early 1990s.[33] Tensions among business leaders, housing activists, and planners grew as they struggled to plan for the future while maintaining links to the unique neighborhood-based politics of the past. In the 1970s, neighborhood activists and a variety of concerned citizens tried to expand the range of public spaces and nascent sustainability politics. By the end of the century, in an increasingly privatized landscape of consumption, and with a lack of consensus, many of the more ambitious goals of Model Cities activists and the counterculture faded. As illustrated by the debate over an ambitious project called the Seattle Commons in the south Lake Union part of the city, during this period Seattle became a model of "sustainable" planning and cultural vibrancy while the unrealized or contradicted goals of earlier activists, who sought to balance equity, environment, and economy, haunted the city's public process amid the latest boom.[34]

Throughout the second half of the twentieth century, cities provided the critical stage on which the drama of environmentalism took hold in the American experience. The WTO protest helped reignite—perhaps only for a brief moment—the global phase of a movement that has, at least in the United States, been inextricable from the changing urban situation over the last four decades and maybe longer. Writing soon after events in Seattle, one prominent activist explained that the WTO protest was not simply "Velcro to which anxious people attached themselves" and that "environmentalists were not just rallying for environmental rights; labor unions were not just marching to protest the disregard of human rights." Rather, he said, the "protesters in Seattle were concerned about the well-being of the world on the whole."[35] This fight for the "well-being of the world" is still, ultimately, a local struggle solved by local people who have an intimate knowledge of what is at stake in their lived environment. But this awareness need not lead to the global version of NIMBYism or the comfortable slumber of insularity that WTO protesters appeared to shatter. Seattleites began to hone this more encompassing sensibility earlier in the twentieth century, and that history—with its enduring conflicts and partial resolutions—if embraced, holds clues that will help navigate the waters ahead.

1

Market

In the summer of 1963 the Seattle Art Museum hung a show by the painter Mark Tobey. A local art world hero, Tobey spent his formative years, during the 1940s and 1950s, as part of the "Northwest Mystic" school, a regional group of painters that included Kenneth Callahan and Morris Graves. By the 1960s his work had garnered national attention, and he had immense cultural cachet in the provincial upstart city. The museum show in 1963—entitled "Mark Tobey and the Seattle Public Market"—presented an artistic intervention in civic affairs, a painted argument about the city that had nurtured Tobey and that he now saw in peril. Tobey dedicated the show's catalog to those "who live in awareness of man's relationship to nature, and who cherish the values of the past as a vital part of the present." Produced between 1939 and 1943, the exhibited paintings and sketches depicted the Pike Place Market as a fecund blend of nature, commerce, and earthy democracy. His canvases captured a built environment teeming with organic energy. In a 1942 painting titled *E Pluribus Unum,* the frame is packed with a sea of human faces, signs, and piles of fruits and vegetables—all the weltering diversity he saw at the market in the early 1940s. Tobey's market was a living organism, a breathing mass of humanity in an idealized civic space.[1]

Despite its recent vibrancy, though, Tobey and his allies in the Seattle arts community saw the market as part of an endangered species in 1963.

E Pluribus Unum, Mark Tobey (American, 1890–1976), 1942. Tempera on paper mounted on paperboard, 19 ¾ x 271/4 in. (50.2 x 69.2 cm). Seattle Art Museum, Gift of Mrs. Tomas D. Stimson, 43.33. © 2010 Estate of Mark Tobey/Artists Rights Society (ARS), New York.

For Tobey and a group calling itself the "Friends of the Market," the old market symbolized what was quickly being lost in the relentless drive toward modernization in Seattle and its immediate hinterland. For the exhibit catalog Tobey wrote, "There is a need to speak, today, when drastic changes are going on all around us. Our homes are in the path of freeways; old landmarks, many of rare beauty, are sacrificed to the urge to get somewhere in a hurry; and when it is all over Progress reigns, queen of hollow streets shadowed by monumental towers left behind by giants to whom the intimacy of living is of no importance. . . . And now this unique market is in danger of being modernized like so much processed cheese."[2]

Ben Ehrlichman, the Seattle Chamber of Commerce president in the early 1960s and an advocate of downtown revitalization, epitomized those who would make processed cheese of the market. For him, modernization was not such a bad word. In fact, he had dedicated his life to building a modern Seattle and to the era's urban renewal. In 1963 the Central Association of Seattle (CA), a group of downtown businesspeople, financiers, and developers, flush from the success of their space age–themed Century 21 World's Fair a year earlier, turned their skills and attention to remodeling the downtown core of Seattle. They were artists too, of a sort. The entire city was their canvas. In 1963 Ehrlichman and his forward-looking comrades in the Central Association and the Seattle Planning Commission unveiled their own masterpiece, an image as sleek and streamlined as the Boeing jets that were then putting the city on the map. Their plan for downtown, as depicted by a meticulously constructed scale model, envisioned efficient paths for consumers in the downtown core, emphasizing "improved accessibility," "major parking garages," and even escalators to ease tired shoppers up Seattle's notoriously steep hills. They argued for mass-transit funding, ring highways, waterfront development, and the closure of downtown streets for a mall system and a downtown shopping center. But the Pike Place Plaza redevelopment project was the centerpiece of their plans. Unveiled to the public several months before Tobey's art show and images of it widely disseminated in the local press, the scale model represented condominiums, expansive modern plazas, and new retail buildings.[3]

Members of the CA had their own ideas about urban sustainability, ones that pertained to business. The 1962 fair tested their vision for the region—efficient monorail transportation, soaring singular modern architecture, and a perfect mix of private and federal funding for development. In significant ways, this clean, modern vision sought liberation from the natural setting and a break from associations with the city's past that Tobey so appreciated. With the new Central Freeway under construction throughout the city and high-rises changing the skyline, Ehrlichman and the CA had concrete reasons to be optimistic about its promises of progress.[4] Hoping to capitalize on the region's recent fortune, Ehrlichman and the downtown business community had planned to redevelop vast stretches of Seattle's urban core beginning in the late 1950s. The only problem was that the new plan would eclipse Tobey's beloved market. The old market they saw as a blighted firetrap ripe for renewal; the market's champions, as misguided and nostalgic.

With local capital and the federal promise of urban renewal subsidies combining to shape and then reshape the nation's postwar metropolitan landscapes, in Seattle the old interface of rural and urban—the molder-

Official poster, Seattle World's Fair, 1962. *Seattle Post-Intelligencer* Collection, image no. 1965.3598.26.91. Courtesy Museum of History and Industry.

ing farmer's market—reemerged in the midst of the "Boeing boom" of the 1960s to take center stage in a public debate about the city's future. Individuals such as Tobey and the grassroots movement that rose up to challenge developers' plans saw preserving the public market as the main front in a larger battle over the urban form, its meaning, and most important, its democratic function. The brewing market conflict arose as middle-class Seattleites stood at the brink of a postindustrial, affluent future that they still imagined lay before them in the early 1960s. These opposing vi-

19

Proposed market development, Pike Plaza Redevelopment Project, architectural model, July 1968. Pike Place Market Visual Images and Audiotapes, item no. 33342. Courtesy Seattle Municipal Archives.

sions of the market suggest the important early outlines of a public discussion about the nature of a "sustainable" city.[5]

On the surface the clash over the market appeared to be yet another fight over the preservation of a historic site during an era of many such conflicts in urban America. In fact, however, the market debate emerges as much more than an architectural preservation battle or simply a clash of two sets of urban elites and their aesthetic tastes. In fact, the market standoff was an early example of an environmental battle and a precursor to the urban environmental politics that would dominate the city and the country after the 1960s. This fight over the "heart of the city" turned on environmental concerns: how unprecedented levels of mass consumption and prosperity for some were reshaping urban and rural places after

World War II, who ought to have the power to decide such changes, and what costs these changes would have for different groups of people in the city and for the city's and its citizens' connections to both built and natural environments. This story concerned how consumption came to the center of urban and environmental politics, as well as how citizenship would be defined and articulated. If the urban environment was in danger of becoming "so much processed cheese," then so were its citizens.

The struggle over the Pike Place Market, although led by contending groups of economic and cultural elites in the city, constituted the first high-profile postwar "citizens'" protest in Seattle, one that spoke directly to a building unease about the future of the urban environment in the United States. The moment of decision for the market—whether to "save it" or "develop it"—appeared with the culmination of postwar planning momentum and the high hopes of downtown boosters. But the character of this contest, especially the focus on living, breathing urban environments, prefigured the neighborhood and environmental activism that would preoccupy diverse communities both throughout the country and in Seattle after the 1960s. The fight over the market, then, was emblematic of a larger American story about the epochal transition from the high-modern city to the postindustrial, service economy city and the incipient shift from an assumed postwar liberal consensus to an increasingly diverse set of activist and grassroots movements animating urban politics. The well-publicized and evolving market controversy distilled growing concerns about history, aesthetics, consumption, social justice, and the environment in one physical place.

A touchstone during a decade of tumultuous changes, the market made the often obscured relationships that linked the city to its environment tangible through the daily, intimate acts of buying and eating.[6] In other words, consumers began to link nature to the urban body—and the citizen body—in a dynamic picture of urban ecology at the market. Consumption became part of an expression of a new politics. A distinct urban environmentalism therefore took hold in Seattle in the midst of this national and local flux. These were the connections of place, memory, and politics that Mark Tobey attempted to manifest in his paintings. They were also the connections that downtown planners such as Ehrlichman would inadvertently lay bare with their own version of an urban utopia, which threatened to erase those connections.

Both developers and preservationists at the time spoke of the market as the city's "heart." In their rhetoric, the market became a metaphor for a package of ideas and a tangible experience that they each thought could define democratic citizenship in the postwar period. Both markets, then—the more abstract market encompassed the exchange of goods and services and that linked consumption to the creation of new metropolitan spaces, as well as the physical buildings of the Pike Place Market—serve as useful starting points for understanding the origins of the postwar urban environmental movement and the roots of sustainability in Seattle.

The Prewar Market and Seattle's Hinterland

The Kent Valley, south of Seattle, may seem to be a strange starting place for a story about the heart of the city after the 1960s. But the valley once fed the city, and the region's metropolitan development—and the transformed producer and consumer landscapes—were most obvious there. Once a highly productive agricultural corridor, the valley stretches thirty-four miles just south of Seattle. Before they were dammed or straightened for flood control and industrial use, the Green and Duwamish rivers had for centuries deposited rich silts in this valley as they wound down from the Cascade Mountains toward Puget Sound. The now traffic-choked landscape once boasted some of the richest and most productive farmland in the United States. Between the turn of the century and the 1940s, the valley was better known for its famous annual lettuce festival than for its sprawl. But tidy rows of lettuce and beets have long since given way to parking lots and a new pattern of the low-slung light-industrial businesses, high-tension power lines, warehouse stores, and tract homes that began to fill the valley after 1960. This is a familiar scene on the outskirts of many cities in the postwar United States. The market once provided an obvious link to the pre-1960s landscape, just as the city's consumers once made farming viable there.

In an earlier era, the market's connection to its hinterland was more direct, and its stalls reflected the diversity of city and rural life depicted in Tobey's paintings. After the 1920s, Italian and Japanese immigrants in particular settled in the lush valleys near the city, carving out a productive and fertile niche near the towns of Auburn, Kent, and South Park. For example, by 1939 Japanese American farmers in Kent and the larger region around Seattle numbered 4,165 and operated at least 476 farms. Despite racist land laws of the 1920s and other efforts to exclude them from the Northwest's farming economy before the war, these farmers, along with Filipino, Italian, and other farmers, provided an essential local food supply for Seattle. Farmers shipped their produce as far away as Minneapolis and Chicago, but the Pike Place Market in downtown Seattle was their most important destination. Today it has become a tourist attraction, a nostalgic evocation of the region's past, yet in the early twentieth century the market was serious business. In 1907 a Progressive-minded mayor of Seattle created it as an experiment, an effort to cut out the intermediaries in the city's produce business who, according to advocates of the plan, were gouging consumers and farmers alike. One of the institution's new rules was an ordinance requiring that only "food and food products raised, produced or made by the person offering the same for sale" be allowed at the market. To take advantage of this new marketplace, farmers would travel long distances: Scandinavians on the Olympic Peninsula might rise at three o'clock in the morning to take the weekly ferry to Seattle. Italian and Japanese farmers from the valley made an easier trip, leaving their horses at nearby downtown stables while they sold piles of produce from their wagons.[7]

The market, which began simply as a designated spot on Western Avenue overlooking the waterfront, soon blossomed into a popular institution and set of permanent structures. With municipal support in the 1910s and 1920s, city engineers drew up plans to build stalls and awnings. Later plans from the 1920s included areas designated for specific businesses, including the first coffee shop and a space for selling doughnuts. The new construction responded to the need to accommodate the automobile, and the interwar boom at the market could be attributed in part to the increased speed and ease with which truck farmers could deliver their produce. As a result, the market saw its peak after 1920, with over 627 permits for sellers. During these productive times and continuing into the Great Depression, Italian and Nikkei farmers from the valley became the main business rivals at the market. By 1940, 60–80 percent of "wet-stall," or lower stall, vendors were Japanese Americans. This was the market that Tobey depicted, a diverse, contested, and busy center of exchange.[8]

The war marked the beginning of the end for the agricultural economy of the Kent Valley, and the market's initial success was also short-lived. Soon after the war began, the incredible success of Nikkei farming operations came to a screeching halt, and so did most transactions at the mar-

Corner of Pike Street, Pike Place, and First Avenue (before construction of the Corner Market Building), 1910. Pike Place Market Visual Images and Audiotapes, item no. 33275. Courtesy Seattle Municipal Archives.

ket.[9] As many farmers remembered bitterly, in May 1942, just weeks before the spring harvest, Japanese Americans were forced onto train cars and sent to internment camps under Executive Order 9066. Few would ever return to farm the same lands around Kent or Auburn. The valley continued to be productive, but it would never again be the same for any farmer. The Pike Place Market, too, would not be the same. The Nikkei farmers were wrenched from their operations in haste; one farmer remembered, "We had to let go of everything."[10] On their way to internment camps such as Tule Lake and Camp Minidoka, Nikkei farmers were forced to leave their ripening produce on the vine and sell their tools and future harvests at rock-bottom prices. In 1939 the market had sold permits for 515 sellers. By

Street stalls on Pike Place, Main Arcade, 1912. Pike Place Market Visual Images and Audio-tapes, item no. 32338. Courtesy Seattle Municipal Archives.

1943 only 196 permits were sold.[11] War left the market experiment indeterminately dormant. When some Nikkei farmers attempted to pick up the pieces of their prewar lives after 1945, they remembered their white neighbors saying, "We don't want those Japs back."[12] Indeed, after the war local anti-Japanese organizations centered in places such as Auburn attempted to drive returning Nikkei farmers from the valley altogether.[13] And not long after, according to one farmer who stayed after the 1950s, "Safeway quit buying local," and "Boeing took over" the valley, dredging the river and covering fertile fields with sand for the foundations of a new, industrial landscape. "Our farm," one erstwhile farmer lamented, "is just a memory. . . . it's all industry out there now."[14]

View of back of Pike Place Market from Alaskan Way Viaduct, Dec. 6, 1951. Engineering Department Photographic Negatives, item no. 43524. Courtesy Seattle Municipal Archives.

The wartime experience of Japanese Americans foreshadowed the abrupt dislocation that would soon follow for many farmers. After the war, the Puget Sound area joined the nationwide rush to "modernize." The Kent Valley epitomized the shifting ground of the regional economy and development priorities. The area south of the city felt these changes acutely. Yet Seattle matched a national pattern as federal policies—especially the Interstate Defense Highway Act, federal mortgage insurance, and urban renewal—continued to stimulate suburbanization of middle-class white people, businesses, and jobs; as large-scale freeway construction physically transformed urban and rural spaces; and as local developers, lenders, and individuals made decisions about using public and private capital to refashion the region.[15] Middle-class suburbanites and large corporations such as Boeing both made out especially well. Boeing saw steady growth in airplane manufacturing after the war and expanded manufacturing jobs between 1949 and 1958. The company directly benefited from wartime contracts for B-47s and B-52s and, in the 1960s, from 727s. Suburban and small-town governments welcomed Boeing's plans for expansion beyond the city limits of Seattle, as well as the residential development that accompanied it. The historian Lizabeth Cohen calls this era the beginning of the "consumer's republic" and notes that at this time the "suburban home itself became the . . . quintessential mass consumer commodity, capable of fueling the fires of the postwar economy."[16] As the aging neighborhoods of Seattle swelled with returning veterans, former farmlands on the edges of the city to the east and west quickly took on a new identity as suburbs: Bellevue, Lynwood, Shoreline, and the like. With the city's tax base moving down the road, urban and suburban policies increasingly left only scraps to people of color, including Japanese Americans, and poor people.[17]

The Kent Valley also reflected national trends in industrial agriculture after World War II. The produce that people consumed after the war came from more distant and consolidated farming operations, such as California's. Part of national trends, King County farms began to vanish, falling from 6,495 farms in 1944 to only 1,825 farms by 1964. The triumph of California's factory farming, long-range trucking, and the consolidation of the farming industry during the war also contributed to the valley's decline. Simultaneously, zoning changes dramatically increased the cost of farming throughout the country, including the Kent Valley. In 1944 the average value per acre for a farm was $290; by 1964 it would be $1,340.[18] In this shift, the fate of farmers in the valley and the health of the Pike Place Market—and the city—were inextricably linked.

During and after World War II, many white farmers left for better-paying jobs and opportunities in the shipyards and factories in the Seattle area and beyond. Some found jobs with Boeing airplane manufacturing, which was quickly growing to become the driving force of Seattle's postwar economy.[19] In 1956 the *Seattle Post-Intelligencer* suggested the tectonic changes. The paper called Kent the "Heart of the Valley of Plenty" but intended a novel meaning. The new economic focus was not stacks of produce but real estate and industry. In one issue, the real estate–sponsored Sunday pictorial section presented a two-page spread with a panoramic photograph of the valley in transition. "Farming, New Industries, and Progress Bless her," wrote the reporter, who explained that such a portrayal of Kent's attributes could not be "passed off" as mere "Chamber of Commerce puffs."[20] Along with presenting its bird's-eye view of the evolving grid around Kent, including new schools, the new East Valley Highway, and the Northern Pacific Railway tracks, the paper emphasized the industrial expansion into the area while nevertheless stressing the productive farmlands as a "cornucopia." With a population of over 4,000 people where there had been only 3,278 in 1950, the already quickly growing valley was perched on the edge of yet another growth spurt due to industrial and suburban expansion. The city had been "trying to keep up with the huge water supply demanded by industry," according to its mayor, Dave Mooney, helping to draw industries such as furniture builders, airplane-parts manufacturers, and Borden Chemical Company, which was "geared to produce 36 million pounds of formaldehyde annually."[21] Increasingly, farmers in the valley would compete with industry for use of water. With sewer expansion, the tax assessment of farmland and therefore its value also changed. New subdivisions, such as "Kenthurst" and "Lake Fenniwick tracts," many of them with covenants requiring that houses "not be sold to, owned by, leased to or occupied by any person other than the Caucasian race," began to fill up the hillsides around the valley.[22] The new developments also mirrored the hardened lines of residential segregation in crowded Seattle.

Such places would make Kent "a Mecca of the city worker," according to boosters, and many had already found positions with Kent's largest employer, Boeing, which hired over 1,100 workers from the area.[23] By the 1950s the valley that once fed the city now fed the world's need for airplanes. The company, with the help of large federal investment in the country's cold war struggle for supremacy in space, quickly made the transition from a wartime producer to a major player in the emerging aerospace industry with both commercial airline business and plum government contracts. The company had already begun its industrial expansion in the suburbs around Seattle in the years during the war. As the less booster-ish *Seattle Argus* newspaper put it in 1963, the company had become "a quasi-public entity" existing, "in large part, by the tax dollars of US citizens."[24] As the federally subsidized engine of economic growth in the region, it exerted an influence that rippled out in every direction. In 1953 the company's investment in plants and facilities was about $60 million, and by 1963, a decade later, with a new dam built to supply water, energy, and flood control in the valley, it had grown to $300 million. In 1962 the company made its largest annual expenditure to date for facilities, most of which were in the Seattle area.[25] It was no wonder that the local press, business community, and surrounding towns watched Boeing carefully for signs of economic health or change during the 1960s. "As goes Boeing, so goes Seattle" was the common wisdom of Seattleites and the local press.[26] But the same could be said for the little towns of Kent Valley. Boeing drove the regional economy and stimulated other small businesses and light-industrial operations that sought to locate near its manufacturing plants in south Seattle, in the suburbs and towns of Renton and Everett, and after 1963, in the Kent Valley.[27]

Boeing and government investment gave shape to the larger transformation of the northwestern economy after the war, setting in place the research and technology infrastructure that in the following decades would drive the next boom, in the information economy, with firms such as Microsoft and Amazon.com. In the valley, however, this transformation required substantial energy, water, and land, making an increasingly hostile environment for area farmers. With federal money and federal contracts, Boeing began remaking its piece of the valley with far-reaching effects for the Pike Place Market and the region. Development of the 320-acre site required, according to the *Seattle Post-Intelligencer,* that the land be razed and leveled by "dumping 500,000 yards of fill material."[28] Much of the financial support for transforming the valley environment came directly or indirectly from the federal government and funded projects that reached beyond the immediate site by raising tax assessments on land, using water, and remaking the physical landscape of roads, culverts, and parking lots in ways that made farming harder. In May 1966, the Economic Development Administration, in the U.S. Department of Commerce, gave the city and county a $5 million grant to "provide highways and sewer and water service needed" for the expansion. Kent and the county then matched the grant to bring combined local and national spending on the valley's remaking to nearly $9.82 million. Construction included a water intake facility, reservoirs, and "transmission mains to the industrial area"; new "sanitary sewers and pumping stations" and a sewage plant; and a widening of area streets and new roads to connect the facility with the existing grid.[29] Senator Warren G. Magnuson hailed the decision to subsidize more infrastructure as a "breakthrough for the industrial development of South King

29

County."[30] As the years passed, politicians and local leaders would drop the pretense that such changes to infrastructure would benefit farmers, focusing instead on the headlong rush to develop a new and more lucrative industrial corridor.

As early as 1965, a slightly wistful and even nostalgic tone entered the boosterism found in local newspapers. In a piece he wrote for his Sunday *Seattle Times* column and titled "Now It's Green (back) River Valley," Dick Moody wrote of "a land rush where bundles of greenbacks talk big and flat industrial sites are coveted prizes has rolled over the Green River Valley community of Kent and on into Auburn."[31] According to Moody, Boeing's decision to build the site in the valley had "triggered a land-buying stampede across the truck farms of cabbages, rhubarb, beans, and berries." One pair of farmers with land near the East Valley Highway whom Moody interviewed described a property tax increase from $25 an acre to $70 projected for 1966. "When we contested it, we were turned down," one of them said, adding that they were told, "You're supposed to put the land to its best use—and farming is not the best use."[32] Farmers who attempted to remain in the valley were thus taxed out of their rural holdings, the more lucrative industrial uses for the land raising tax assessments beyond their ability to pay. Boeing's success would spur a variety of suburban industrial expansion into nearby areas such as South Park and other former small farming towns.[33] By 1968 sharper critics of these changes, such as the *Seattle Argus,* described with more outrage the inflated land prices in the valley and the recently completed "high-speed freeway slashing through the heart of the verdant Green River Valley." The new freeway, the newspaper argued, had brought "people speeding into Kent, along with new industry, stores, schools, and problems."[34]

Consumers and consumption were an important factor in reshaping both urban and suburban environments well before the 1960s, but the true driving force consisted of large property owners in the valley, developers, and businesspeople in Kent and Seattle. The choice to invest in infrastructure and exclusive suburban neighborhoods contributed to a larger national trend of investing in a consumer landscape that relocated what had once been vital urban functions to the periphery of cities and into land that had once produced food for cities' markets.[35] At the same time, these investments left people of color behind. As U.S. economic health increasingly translated into maintaining high levels of mass consumption, the consumer—or an imagined consumer—became a powerful and volatile figure in the process of urban and environmental changes. When downtown developers in Seattle watched apprehensively as these consumers began "speeding into Kent" and other suburbs removed from the older core, they devised ways of courting them. For people in the city who paid attention, the market came to symbolize a kind of indicator species in this quickly changing metropolitan environment. By the 1960s the deteriorating and increasingly dilapidated buildings of the market suggested a link that had been broken.

Making a Metropolitan Landscape

Men such as Ben Ehrlichman had interests that straddled both urban and suburban spaces. Ehrlichman, a prominent investment banker and developer of many of Seattle's significant structures in the downtown core between 1930 and the 1960s (including the Exchange Building and the Century 21 fair constructions), invested in suburban shopping malls as well as downtown. He was one of the principal planners of Northgate Mall, just north of Seattle near the Interstate 5 freeway, one of the first indoor shopping malls in the United States, which opened in 1950.[36] As in other West Coast cities, downtown department stores such as Bon Marché and J. C. Penney built anchor stores in suburban shopping centers to cover their bets. Ehrlichman and others of his ilk hoped to reap benefits from the shifting geography of postwar development, yet privately they were also greatly concerned about making the most of these economic changes for the entire city. They grappled in their own way with the downtown environment's incipient decline. To coordinate efforts of downtown renewal, Seattle's Chamber of Commerce, leading downtown developers, the publisher of the *Seattle Times*, and architects organized themselves to plan what they believed would be a more viable and accessible city for people who increasingly tended to their business in the new suburban malls and who worked and lived beyond the city limits.[37]

An image of the ideal consumer and an ideal consumer-oriented urban landscape formed an explicit focus for the CA's energy. Between the mid-1950s and the 1970s, civic leaders, especially downtown business owners, reacted to the new postwar circumstances in an increasingly coordinated fashion. As Seattle's downtown declined, these civic leaders turned to a set of principles derived from the experience of other cities during the

same period.[38] With Boeing leading the way for high-tech industrial development after the war and regional boosters celebrating economic and demographic expansion in the region, Seattle's downtown businesspeople
were cautiously optimistic about the city's future.[39]

The downtown business community first offered a glimpse of its hope
for the future in its plans to make the core as accessible as a suburban
shopping mall. Freeways were crucial to knitting together this new landscape. Seattleites needed only look to the hills near Lake Union or to
swaths of destruction north and south of the city, after 1958, to see the un-
deniable changes that freeways would soon bring to the city. That year,
the Washington State Highway Commission began the methodical work
of evacuating Seattle homeowners from the path of the new Interstate 5
freeway then under construction. These homeowners felt intimately the
emerging consumer landscape that Ehrlichman hoped to shape. In an article titled "Tornado's Path? No—It's the Freeway Route," the *Seattle Times*
chronicled the contemporary scene: "Concrete basements and house
foundations are all that remain of the residential area in the foreground
of this photograph of the Seattle Freeway route looking south from Sixth
Avenue Northeast and 43rd Street across the Lake Washington Ship Canal toward Capitol Hill." For miles to the north and south, it added, "condemned houses in various stages of being moved or wrecked line the freeway route."[40]

For downtown business owners and developers, freeways presented
both opportunity and potential peril. Earlier federally subsidized road
building had already begun to drain people and business investment from
the city before the construction of the Central Freeway project began. Indeed, the downtown business community formed the Central Association
of Seattle in the late 1950s in an effort to define their interests in the face
of the "urban flight" during the Boeing boom.[41] The first preview of its coordinated plans, a joint policy statement published with the city planning
office in 1959, displayed the harmonious relationship between and common goals of business and government that prevailed in the city during
the early 1960s. The report envisioned a metropolitan future in which "a
healthy downtown serves and supports the surrounding business and residential communities just as they, in turn, complement and use the Central
Area." But to ensure that this harmony continued, the CA proposed "more
orderly and efficient development" using "private and public investment."

Freeway construction near the intersection of Lakeview Boulevard and Eastlake Avenue, 1961. *Seattle Post-Intelligencer* Collection, image no. 1986.5.4016.1. Courtesy Museum of History and Industry.

In its extended argument for downtown's existence, the CA presented a forward-looking bill of goods that emphasized rapid transit, better parking facilities, and most significantly for the Pike Place Market, "redevelopment of older blighted neighborhoods." The developers offered their own consumer-driven vision for a sustainable urban center in a period of flux. They argued that the plans would return such property to "its highest and best use" and promoted an ambitious set of freeway construction projects to complement the new Central Freeway, including a highway ring around the city and an "internal circulation system" that would make the core more accessible. The early report sketched out the basic set of ideas that would dominate urban priorities for the next decade.[42]

When the *Seattle Times* published a "progress report" (created by the

33

Central Association) in 1961, it provided another of the earliest publicly available glimpses into the downtown business community's determination to rebuild downtown while retaining what it called a "Downtown for People." This special Sunday supplement emphasized consumption and accessibility, featuring happy shoppers moving across an ultramodern sky bridge that connected Seattle's venerable 1920s Bon Marché department store with a new parking structure across the street. Ads for the store boasted the "self parking garage" and "Seattle's Downtown Drive-In Department Store . . . the most modern shopping convenience in the world." Other companies, such as Washington Securities, advertised "double the parking facilities"[43] and featured pictures of garages in their advertisements.[44] In the article the Central Association promised to publish a report the following year revealing how it would accomplish the goal of making Seattle more than "just a black dot on a free road map" and realizing its "promise as the dominant city of the Pacific Northwest."[45]

In 1963, two years after this "progress report," the CA and the municipal planning commission officially published their completed plans for remaking downtown in the image of a suburban shopping mall.[46] The *Seattle Times'* preview suggested that accommodating the car and making the core accessible would also mean drastically remaking the character of downtown Seattle, a profound transformation of the core business district into a "mall system" that would eventually close the east-west thoroughfares Pike and Pine streets to "tie together the major retailing areas" and "provide pedestrian access to a revitalized Pike Place Market area." They proposed "canopies over sidewalks" to "provide all-weather protection" for the less-hearty Seattle shoppers, along with escalators that would "solve the problem of steep grades," "serve as connectors with major parking areas," and "be tied in with transit stops."[47] Just a year after the Century 21 World's Fair, the Seattle business community hoped to take the lessons they had learned on the utopian fairgrounds—home to the Space Needle—and apply them to the core. "The city that surprised America and the world by staging the first profitable World's Fair in the nation's history isn't resting on its oars," the CA boasted. Its new central district plan "aimed at remolding the downtown area."[48]

In this view, the Pike Place Market's dilapidated structures were a source for renewing the downtown's fortunes rather than a symbol of loss. The Pike Place Market, the center of this "remolding," formed the focus of the Central Association's comprehensive plans. If the market's decline indicated any failures or uneven development in the evolving landscape of metropolitan Seattle, the CA saw urban renewal as the remedy. Describing the market area as "a new 'front door' for downtown,"[49] the CA proposed its "Pike Plaza Redevelopment Project" plan, which would take an area that it described as the "most dramatic downtown setting to be found in any city" and remake it as a "vital living-shopping-recreation center, retaining the public market as the nucleus" around which redevelopment would occur.[50] In the initial report, the site included twelve acres, and its transformation was to constitute a "major Downtown development." The plan suggested the market be "rebuilt as a major visitor attraction" sitting atop a generous new 3,000 car-parking facility with a park area and "ample space for important new structures, including a hotel, shops, and a limited number of high-rise apartments to the north."

Ironically, though the Pike Place Market had fallen victim to suburban growth and the new postwar consumer economy, downtown developers now hoped to remake it to serve the new idealized consumer. And in a brief but important line in the report, the authors noted that the Pike Place Market redevelopment "could be Seattle's first major Downtown Urban Renewal project."[51] At the height of the postwar Boeing boom in Seattle, the business community was set on seizing the moment when national spending on urban areas was at a high-water mark. Defining the market as a blighted neighborhood in need of renewal—even if their intent was to raze the entire neighborhood—assured them a piece of the federal pie.

Although many neighborhoods and sections of the city could have been considered blighted and certainly underserved, obtaining federal urban renewal funds required heightening the image of a downtown business district in crisis. A year later the Central Association and a newly formed entity called the "Seattle Urban Renewal Enterprise" unveiled their "Pike Plaza Redevelopment: A Preliminary Feasibility Study," a plan that turned on using urban renewal funds. Supporters and the study committee included a who's who of downtown businesspeople, pillars of the community, and members of the Central Association who argued that buildings in the market area were "generally blighted and, undoubtedly, would qualify under both State and Federal laws for redevelopment by Urban Renewal." The report claimed, in stark contrast to Mark Tobey's idealized image, that the market presented "potential hazards to public health, welfare, and safety" and contributed to the "spreading of blight," thus having a "debilitating effect on the Central Business District."[52]

This fear of creeping blight suffused the report's twelve main conclusions and recommendations, which consequently suggested that "the power granted by the State Legislature to assemble blighted properties by use of eminent domain" be used to "redevelop said properties and make them available for private development." To meet standards under the Urban Renewal Act—an act that was originally drafted to alleviate overcrowding and unsafe conditions in segregated neighborhoods such as Seattle's Central District—the market had to be redefined before it could be remade with federal funds into upscale condominiums and a shopping mall. Adding insult to the injury of this disingenuous plan, in 1964 Seattle voters turned down a referendum for open housing after years of efforts by local civil rights groups. The willingness to work fast and loose with urban renewal money—bulldozing a neighborhood to build luxury condominiums—only underscored this neglect of the city's truly troubled neighborhoods suffering from blight and crowding exacerbated by segregation in housing and the workplace. The report, though, clearly revealed this blind spot, as well as the frustration among downtown developers that such a prime piece of real estate had been left untouched for so long. Blight was their argument. In the drive to modernize downtown, the market represented all that these developers hoped to "modernize," and according to the plan's authors, its remaking would be "the single most important redevelopment project in Seattle's history."[53]

A mere fifteen years after the war, according to the business-centric narrative, Seattle had transformed from a small urban center with a regional economy based on resource extraction and dependent on its immediate hinterlands for food to a leader in the postwar, high-tech consumer and military-industrial economy. Shedding its provincialism to become a "great metropolitan area," as Ehrlichman and the boosters hoped, also meant disregarding the recent human and natural history of places such as the market and Kent Valley. The same desires that drove suburban development and rearranged the region's economy also struck at the heart of the city's human environments. That same momentous year of big plans and high hopes, near the end of a distinguished career as a committed city builder in the Northwest, Ben Ehrlichman assumed the presidency of the Central Association. In the organization's annual report, he linked Seattle's efforts to a national context, writing, "Property owners, merchants, leaders in every segment of downtown business are working together to rebuild the downtowns of America." The job ahead would require "teamwork" between "private enterprise and government" and "certain calculated risks."[54]

As Ehrlichman knew, creating a new consumer landscape would have serious consequences and therefore would have its critics. The association saw its plans as strengthening the heart of the urban body, a dynamic grab for the future. And Ehrlichman drew from nature for an apt metaphor to illustrate the CA's attitude toward resistance: "Grasp the thistle firmly; hesitate, touch it timidly, a thistle stings. But grasp it firmly, its spines crumble harmlessly in your hand."[55] Addressing the organization at its annual meeting in Seattle's Washington Athletic Club on May 18, 1964, Ehrlichman looked back on their six-year effort to craft comprehensive plans for the city. He told his business associates and the honored guest (the pliant Seattle City Council president) that the "big task" was "to translate these plans into action." "We are now presented with the challenge of rebuilding the very heart of the city," Ehrlichman told the crowd, "and what great metropolitan area can be healthy without a strong heart?"[56]

Resistance at the Heart of the City

The growing "spines" of dissent were various, although the principal authors of this movement's powerful arguments and actions were mainly middle-class women and prominent men in the design, architecture, and arts scenes. As the decade progressed, however, the women and men who would challenge the developers' view of the market were increasingly students, hippies, young professionals or families, and others who nurtured a different consumer vision of the city, its past, and its future. To begin with, many of the market's preservation advocates were concerned about elite aesthetics and good design. But they also simply liked the neighborhood the way that it was, and especially its vernacular qualities. Such concerns were not as simple as a preservation battle to save the market as a piece of architecture. These actors reveled in the existing structures, but they also enjoyed the sights and smells of the place, its atmospheric qualities. They saw it as an organism. The market struggle came to embody a vision of urban ecology that reached beyond the physical structures to the relationships in which the market participated. During a decade of contest and opposition to development, grassroots activists nurtured an alternative form of consumer and democratic citizenship that emphasized the organic, the natural, and the regional.[57]

To be sure, there were leaders in this emerging consensus. An architecture professor at the University of Washington named Victor Steinbrueck gave voice to a persuasive and powerful urban vision for Seattle in the second half of the twentieth century. With his series of books, journalistic articles, and public appearances, Steinbrueck provided intellectual vigor, a public face, and professionalized legitimacy to an urban sensibility that appeared to be growing spontaneously in the early 1960s.[58] A founder of Allied Arts, Steinbrueck, along with other prominent members of the local arts community, hoped to shape the tastes of Seattle and uplift its cultural standards.[59] According to the historian Roger Sale, these new activists and civic stewards were "interested in the arts, in the development or renewal of neighborhoods, in historic preservation, in stopping the construction of new big highways"; they were people for whom, he wrote, "the basis of new possibility was old Seattle, the city that had been built by 1915."[60] Nowhere was this building civic spirit more evident than in the pages of *Seattle Magazine*, where journalists strenuously criticized the new I-5 freeway's "hideous design," suggested "sign control" for "neon eyesores," and described other "roadside abominations."[61] This culture and politics monthly, produced by Seattle's progressive Bullitt family, sought, as Sale argues, "to become the conscience of the city."[62] From 1964 until it folded in 1970, the magazine perfectly reflected the ethos of a new constituency of urban leaders rising to prominence in the mid-1960s, especially with its encompassing concern for the urban environment. Throughout its run, the magazine routinely published articles discussing how to cure the city's ills. In its first issue the magazine included a review of *God's Own Junkyard*, a recently published book influential in the debate over blight in the American landscape. Steinbrueck, the review's author, connected Seattle's plight with the national conservation movement; "Though Seattle escapes specific mention," he wrote, "the attack launched by *God's Own Junkyard* is highly relevant to the Puget Sound area." "Like the rest of the country," he continued, "Seattle is beset by visual pollution and destruction, and prospects for betterment are gloomy." The "flood of ugliness keeps rising," he added, "swept along by the aesthetic indifference" of citizens and leaders.[63]

38

Throughout the 1960s, various civic organizations, the editors of *Seattle Magazine*, and Steinbrueck argued that the "mounting tide of defacement" could be turned only if "local and national leaders" were "prodded by public outrage."[64] The magazine acted as one organ for this new bourgeois outrage in Seattle—a marked departure from the status quo of the Central Association, which they believed had brought acres of gray concrete devastation to the city.

Steinbrueck is remembered mainly as a preservationist concerned with architectural aesthetics. Yet his urban vision reached well beyond architecture or perceptions of urban "defacement." In fact, his attention to both built and natural forms in Seattle created an opening for an urban environmental movement that not only encompassed both but always included a humanist component in its consideration of the city. As the decade wore on, his arguments for preserving the market presaged the concerns of a much broader critique embodied in the popular ecology movement. His efforts in Seattle would draw increasing legitimacy and energy from that movement later in the decade. He invited Seattleites to develop a heightened aesthetic appreciation of their shared vernacular landscape, and in this he was not alone. During the late 1950s and early 1960s, the University of Washington's College of Architecture and Urban Planning and new Landscape Architecture Program nurtured faculty who helped support this new consensus about the urban form. In particular, Meyer R. Wolfe and Ernst Gayden of the Urban Planning Program, and Richard Haag, the founder of the Landscape Architecture Program and designer of the innovative Gas Works Park in Seattle, among many others at the University of Washington (UW), would join the market preservation fight, lobbying the city council.[65] Steinbrueck's technique of representing urban space in images, the architectural historian Jeffrey Ochsner argues, was influenced by the Townscape movement, a critique of modernist urban planning derived from the British critic Gordon Cullen that emphasized the experiential dimension of urban spaces, the idea of sense of place, and the accretion or development of urban spaces over time.[66] Wolfe, Steinbrueck's urban planning colleague at the UW during the 1960s, similarly explored the city as a "physical entity" in works such as *Towns, Time, and Regionalism* and *Uses of the Anachronistic*. And both Gayden and Haag explored the links between ecology and urban planning in their works.[67] These academicians would provide legitimacy and intellectual weight to the market controversy and other urban-design issues of their time, but perhaps more important, they would influence the next generation of young and idealistic urban-design professionals who would become involved with the Model Cities program and creative approaches to planning in the 1970s.

To understand what drove this emerging movement to challenge downtown developers and why the CA developers might have seen the movement as misguided and even incoherent, it helps to remember what downtown and the market were like in the early 1960s and what people like Steinbrueck valued. Ehrlichman and his comrades could be excused for not seeing the positive qualities of the old, dilapidated market. And here, Steinbrueck's careful and honest examination of the Seattle urban form is invaluable for capturing the historical realities of the market in the 1960s.

Steinbrueck never sugarcoated the places he described. The architect and professor cataloged, sketched, and described everyday places downtown—not architectural gems—as he made the argument for a different beating heart of the city. The atmosphere of the market in the early 1960s was rich. Well before the controversy took root, Steinbrueck had consistently illustrated the market's qualities in his early book of sketches and commentary *Seattle Cityscape*. Steinbrueck published his inventory and loving portrait of the city's built and natural landscape in 1962, the same year visitors streamed into Seattle for the world's fair, hoping the book would help "the observer to participate more fully in the experience of knowing and comprehending the city."[68] He later called himself "a graphic environmental propagandist,"[69] and his line drawings and handwritten script embodied a new sensibility with an antitechnocratic flavor, a direct response to the official reports and surveys of the Central Association. His depictions blended the built and natural environments of Seattle, creating a broadly defined portrait of the cultural and physical landscape that could include water, hills, buildings, and bridges. "It is a basic premise of this book," wrote Steinbrueck, "that esthetic experience is a part of everyday life, an essential human value. . . . There is no one who is not affected by the quality of his environment."[70] He complained about recent planning that showed a lack of respect for the natural setting. "Major changes to the contours of our city still show open wounds and scars and cannot be held to be unmitigated blessings and improvements," he pointed out, adding, "Much of 'suburbia' is guilty." With a critique more pointed than that of the emerging freeway protesters, he attacked what he called the "ruthless brutality of the latest freeways, expressways, and their structures in ignoring the qualities of the locations through which they move," labeling such constructions an "obvious example of short-sighted disregard for human and natural values in favor of narrow technical considerations of automobile movement."[71] His arguments in *Seattle Cityscape* emphasized a desire for balance, but Steinbrueck also suggested the conflicts over aesthetics and public space brewing in the city during the early 1960s (he would publish *Seattle Cityscape #2* in the 1970s). Like Jane Jacobs and other postwar critics of freeways and urban renewal, Steinbrueck celebrated a more holistic and scrappy conception of the city and its organic functions.[72]

The Pike Place Market epitomized Steinbrueck's larger urban vision. In the early 1960s the market might have been hard for some people to love, even those who approvingly read his books. Steinbrueck admitted that it was a rundown neighborhood. The urban planners and downtown developers who feared its metastasizing blight might be excused for their judgments. But the market's unsavory qualities were also central to its charm. Steinbrueck was most articulate on this point. Describing First Avenue, the main artery bordering the neighborhood, he wrote in the *Post-Intelligencer* that the area was the "most colorful, vital and interesting of all." Although a committed modernist with a hand in designing the Space Needle, Steinbrueck was also devoted to the vernacular. Like the landscape studies pioneer John Brinkerhoff Jackson, his contemporary, Steinbrueck was drawn to the "stranger's path," the skid rows and seedy areas of the city that catered to the needs of mainly male transient workers. He turned the Central Association's notion of blight on its head, making the seedy market neighborhood the existing "heart of the city." The "problems of First Avenue are the problems of our city and our society and not to be solved by being ignored nor by cosmetic architectural surgery," he argued. Such surgery would sweep away the key features that made the stranger's path so vibrant: "the workers, the sight-seekers . . . the transients, the woodsmen, the sailors, the first Americans and all colors and nationalities." He even celebrated First Avenue's signage: "Paris Book Shop—Novelties—Party Records, Go-Go Girls—Dr. Pepper, Totem Pole Loan, Cave Tavern." Steinbrueck and his followers were more than historic preservationists; they hoped to preserve a whole set of associations. In response to planners who claimed they could replace such a vibrant scene in the middle of the city with a retooled market, Steinbrueck responded that "no architect would pretend to be able to recreate an atmosphere such as this, which has been growing for more than a half a century."[73]

In 1968, during the height of the debate over the market, Steinbrueck published an illustrated book similar to his earlier *Seattle Cityscape* books, but this one, *Market Sketchbook*, focused exclusively on the market. From the brown-paper-wrapper style of its cover to the line drawings and hand-written script, the book revealed Steinbrueck embracing blight and the organic neighborhood. With text and image, Steinbrueck walked his reader through the neighborhood from First Avenue, the "honky-tonk street" he described in 1965, down the hill of Pike Street, and into the heart of the market. His guided-tour coffee-table book targeted a sensitive new audience of urbane Seattleites yet celebrated the area's grunge. Looking down Pike Street, he wrote, one would see "a drug store, a clothes store, an all night motion picture theater, a fresh doughnut shop, one of the most complete hardware stores in Seattle, a pet shop, a passport photographer, amusement arcades, a peep show, a cabaret, a pawn shop, an antique shop, a barber shop, a shoe shine stand, several cafes, several hotels, several taverns, and a tattoo artist." He added, "On a clear day, Mount Rainier can be seen in the distance."[74]

For Steinbrueck the market was more than a set of buildings; it was a set of relationships that evolved over time—an ecology. He used this kind of vocabulary to describe it. Remarking on the number of popular second-hand shops in the market, he described "salvage clothing and surplus stores, taverns, small hotels, and rooming houses" as an "integral part of the ecology of the Market." His sets of drawings included interior images of some of Seattle's oldest residential hotels, and Steinbrueck argued that the "low-income and working class" people who lived there were the lifeblood blood of the place. Similarly, he mapped out the entire market, with inch-by-inch descriptions of all its nooks and crannies. He pointed out the stairway leading to the "only publicly maintained men's lavatory . . . in the central business district." Toward the center of the old market's main arcade, where local produce was "framed and brightened to a glistening sparkle by the hooded overhead floodlights," he remarked, "an inviting slot to the left reveals itself as yet another rampway to the lower floor." At the bottom of the ramp, he wrote, "the space expands into the large cavern of the lower floor" of upholstery shops, malt shops selling beer-making supplies, a small post office, a Goodwill thrift store, and a shoe repair shop, among

Hillclimb Corridor (looking west through upper-level main market breezeway), 1972.
Pike Place Market Visual Images and Audiotapes, item no. 36198. Courtesy Seattle Municipal Archives.

other business and surprises. Throughout the tour, Steinbrueck emphasized the diversity of the inhabitants: Greek and Italian shopkeepers, the
remaining Japanese American produce companies, and other business
owners. Speaking of the famous intersection of First and Pike, "where the
character of the Market begins to be revealed," Steinbrueck summed up
the meaning of the place, its mixture of old and new, poverty and afflu-
ence, and "rusty marquee[s]": "It is a hustling, bustling scene of essential
human activity. The life of the city flows through here. It's not a lonesome
place."[75]

Steinbrueck may have described the place accurately, but he and the
CA saw the same elements quite differently. For the latter, the market and
downtown that Steinbrueck celebrated evoked the dangers and even de-
pravity of old Skid Road and certainly not the heart of the downtown. In

Triangle Building—before rehabilitation (looking north along Pike Place and Post Alley), 1972. Pike Place Market Visual Images and Audiotapes, item no. 34432. Courtesy Seattle Municipal Archives.

the market, money and space were wasted catering to the wrong kind of consumers and the wrong kind of people. If the CA hoped to remake downtown in the image of a suburban shopping mall, and if it hoped to attract white, middle-class women to the city in particular, Steinbrueck's picture of a male-dominated, rough-and-tumble world was unsatisfactory. The important elements in this unfolding drama regarding the urban form would revolve, then, as they had in other cities, around a set of assumptions about what white, middle-class female consumers wanted from downtown in the mid-1960s.[76]

As part of its own decade-long effort to shape downtown, the CA surveyed workers there so as to assess the problem of "decline," a decline that it linked directly to a reduced "use of downtown shopping facilities" and blighted areas such as the market. The ensuing report noted that "businessmen in the downtown area have viewed the trend of moving to the suburbs with alarm, and rightly so," but that "unfortunately, viewing with

Bakery Building, Pike Plaza Redevelopment Project, May 22, 1968. Pike Place Market Visual Images and Audiotapes, item no. 32417. Courtesy Seattle Municipal Archives.

alarm does not solve the problems of not only 'saving' the downtown area, but revitalizing it." The report's authors argued that cities "degenerate" for economic reasons or because of a "change in the habits of the population," which they argued were "intricate and complex."[77] The CA hoped to use downtown workers as a control group to assess these changing habits and "provide information by which the importance of the downtown area as a shopping center may be determined." The results, its members hoped, would "provide a solid foundation for planning the renewal."[78] Despite the scientific conceit, Seattle's planners followed a pattern common in other American cities during the heyday of urban renewal, hoping to make an argument for attracting federal money and to leverage local political will. The survey's results were especially revealing about the preferences of women, one of the ideal consumers that downtown businesspeople supposedly had in mind when establishing a foundation for planning.

Certainly Seattle's downtown developers took the ideal citizen to be a white, middle-class female consumer, someone for whom the older city, they assumed, had become too cumbersome and unattractive. Indeed, the images of Skid Road that Steinbrueck embraced were thought to be the problem. Just as Steinbrueck celebrated a mostly male-dominated landscape and history of work, the CA used assumptions about gender—especially concerning middle-class women's preferences—to justify its plans. The CA's local renewal strategy was thus in concert with the national trend at the time, which was to remake downtowns to attract middle-class female consumers living in suburbs while simultaneously marginalizing people of color.[79]

But such assumptions about women and downtown were not borne out in the study that the CA commissioned. To begin with, women outnumbered male downtown workers by 10 percent overall and in almost every category that the CA devised. These single working women showed a higher level of satisfaction regarding the city as it was. It is also worth noting that women who worked downtown as retail clerks and office workers earned less than men and were approximately 25 percent less likely to be married, according to the report. Women were twice as likely to take the bus and half as likely to drive downtown. A section titled "Preference for Place of Work" summarized preferences about job locations, city or suburb. Approximately 60 percent of the men surveyed stated that they "would prefer the suburbs," while 65 percent "of the female respondents gave downtown as their preference." Women found downtown more satisfactory for shopping, more convenient, and even cleaner than did the men surveyed. And both men and women put transportation and parking at the top of their lists of things to change, rather than removal of tenement houses, the installation of "moving sidewalks," or the need for cleaner streets and shelters, for example.[80]

Women, the ideal consumers that male planners hoped to woo, seemed more than satisfied with downtown the way it was. They were most concerned about transportation, especially bus service. In contrast, men, it seemed, were the most squeamish about downtown. Perhaps, therefore, the report suggested that the CA projected its members' own ideas onto women. In any event, the plans they would push forward in the following years had little to do with what working and middle-class women downtown said about their experience of the city. This misunderstanding would come back to haunt the CA during the debate over the market, for women would form the critical base of grassroots support for preserving the neighborhood.

Before Steinbrueck or *Seattle Magazine* had even begun to articulate their vision of the city, similar sentiments appeared among one group of aesthetically and environmentally minded middle- and upper-class women, the target of the CA's efforts. Steinbrueck and Allied Arts participated in antifreeway protests during this period as well, but women took the central role in the drama and its characterization in the press. They began with a small protest regarding the new I-5 freeway construction through the city in 1961. The protesters hoped that the state would construct a lid over the portion that would run through downtown. "With a four piece band blaring 'When the Saints Go Marching In,'" a crowd of nearly one hundred marchers—"beauty lovers all," according to the *Seattle Times*—"hiked behind a police escort" on a pleasant afternoon along the proposed "freeway ditch" to "express their hope for a lidded park" that would prevent the city from being cut in half. The first note of opposition, then, came not so much from people who were opposed to freeways as from beautifiers who hoped to take the rough edges from the freeway project. In fact, reflecting the findings of the CA's own study of downtown workers' habits, many of the people who would form the growing opposition to the CA's plans at the market were women. Nonetheless, the marchers, who, witnesses said, "clutched sprigs of flowing greenery in their hands" or, as did "many men," "wore greenery in their buttonholes," foreshadowed the Earth Day protests in the decades to follow.

One woman leading the protest told reporters: "To us, this ditch makes no sense. Men can argue all day about the relative merits of blondes versus brunettes versus redheads. But they all will agree they don't want to see a bald-headed woman." The urban beautifiers argued in gendered terms that to build an uncovered freeway through the city was "to go for the highway version of a bald-headed woman."[81] The playful rhetoric of the antiditch group hinted at a much broader movement that would gain momentum by the end of the 1960s and in the following decades.[82] The very consumers whom downtown developers hoped to attract often challenged the new landscape instead, and they used a gendered language to do it. They foreshadowed impending fights over public space at the market and the reassessment of the city's priorities.[83]

When the CA unveiled its new streamlined plans for the Pike Place

Protesters march across Spring Street demonstrating opposition to freeway construction, June 1961. *Seattle Post-Intelligencer* Collection, image no. 198E5.4018.1. Courtesy Museum of History and Industry.

Market, Steinbrueck was the first to petition the city with a thorough critique of the plan, but he was soon joined by this group of activists, who began to define the link between the production of urban space and their political activism. In the statement he submitted to the city council, Steinbrueck lambasted the details of the plan as "cheap" and "unimaginative" and denounced the city planners. "It is an elementary principle of planning in a democratic society that the experts prepare and present alternate proposals for public consideration," he wrote.[84] Steinbrueck made clear

Victor Steinbrueck leading a demonstration against demolition of Pike Place Market, 1971. *Seattle Post-Intelligencer* Collection, image no. 1986.5.54096.1. Courtesy Museum of History and Industry.

Victor Steinbrueck and a young woman protesting the proposed demolition of Pike Place Market, 1971. Photograph by Timothy Eagan. *Seattle Post-Intelligencer* Collection, image no. 1986.5.53773.1. Courtesy Museum of History and Industry.

his opposition to the Pike Plaza plan, in which the market was to be "eliminated and replaced by something else which is called by the same name." Nonetheless, he added, "it is definitely eliminated."[85]

While Steinbrueck's criticism pointed to what he saw as the undemocratic nature of the process and the poor planning, just days later a group of citizens lodged their own complaints. Once again, women were critical voices of resistance, making up approximately one-third of the signatures on the "petition of citizens for maintenance of the present appearance and general character of the Pike Place Public Market." The other dominant group comprised professional men, architects and engineers, all of whom wished "to go on record as being completely opposed to any plan that would move the market from its present setting," as the downtown plans entailed.[86] The following year, a letter-writing campaign—again, predominantly by women—began to pummel the city council with letters urging them to "save the Pike Place Market as is."[87] One letter writer, referring to herself as Mrs. Robert F. Ray, wrote that the market was to Seattle "what Les Halles or the Flea Market is to Paris." She exulted in the "stalls, restaurants and shops tucked away in its dark corners." She further remarked on the "smell of saltwater and fresh fish which gives the Market a true Seattle flavor." On a more practical level, and reflecting the experience of women surveyed for the CA's own study, she found it easier to visit the market: "While attending business college in downtown Seattle I often am unable to reach a supermarket before the meat sections close at 6 o'clock. I find it easier to go to the Pike Street Market for my meat."[88]

The torrent of criticism against the plan began to flow in 1963 and built in the following years as the CA attempted to push it forward with support of a sympathetic mayor and city planning department. By the mid-1960s the arguments against market redevelopment evolved from the generally aesthetic concerns of Tobey and Steinbrueck to include a broader and more open-ended critique of postwar metropolitan change. This critique emphasized links between a democratic society and alternative consumer choices that the market represented during the 1960s.[89] Letter writers, petitioners, and Steinbrueck himself began to incorporate more potent sentiments about democracy, citizenship, diversity, and social justice within what had been merely aesthetic arguments. The evolving opposition to the CA's development schemes reflected clear links to a broader and shifting context of urban politics during the civil rights era, as well as showing the mix of urban and environmental politics. Steinbrueck and his Friends of the Market, as well as letter writers in the second half of the 1960s, tapped into the contemporary debates about slum clearance and misused urban renewal funds in other cities. And although led by many of Seattle's new arts and culture elites, the appeals against market redevelopment were distinctly populist, reflecting or perhaps prefiguring the shifting ground of 1960s politics, including the postwar women's movement.[90]

Steinbrueck's National Register nomination for the market in 1968 reflected the way most activists had begun to understand the market's significance by the late 1960s—as a diverse and living organism, like that depicted in Tobey's artwork. The architect and activist called the place a "unique living heritage" and a "sociological mixture of all people." Echoing Mrs. Ray's letter, he emphasized "the process of food preparation and the availability of local produce sold by the farmers who grow it" as providing an educational experience. The architecture itself, he wrote, was constantly modified by the "varied display of food and objects accentuated by sounds and smells," as well as "passage ways and openings with varied spaces, shapes, stalls, ramps." He emphasized what he called the "dramatic experience of people acting out their daily existence through face to face involvement, in contrast to the sterile, dehumanizing environment that has grown to be typical of much of our urban world."[91]

Whereas the Central Association hoped to create the clean lines of the modern city, the preservation advocates understood the Pike Place Market as far more than architecture. One letter writer confirmed the mood of many in the city, comparing the market battle to other environmental issues: "Each week one reads of the necessity of the formation of a new group of private citizens to fight the battles to 'save the Market,' 'save the Arboretum,' 'save Lake Union,' or 'save Lake Washington.'" He asked why the city council or the legislature could not do its job to "look after the welfare of the entire populace." For many activists, the market did not just represent diversity but offered an antidote to an increasingly sterile and inorganic urban and hinterland environment. One writer went as far as comparing the intricate spaces of the market to "rapidly disappearing clean air and water" and "great open space."[92]

In March 1969, just months before the city council was scheduled to decide whether the city would proceed with the Pike Place redevelopment plan, Steinbrueck led a rally of more than three hundred people at the Moore Theatre, near the market, and on March 19 he led supporters to city hall. In the several months leading up to the crucial vote and council meeting, a flood of petitions against the city plans once again made the link between the market and a larger landscape of social change. Steinbrueck was working within a context of heightened racial tensions in Seattle and throughout the country by the late 1960s. He led the protests alongside a highly mobilized student movement against the war in Vietnam. The downtown business community had begun to take notice of such activism by 1969 as more and more students moved their protests from the University District to downtown and as a variety of citizens' groups tackled urban issues. The people who marched with Steinbrueck could no longer be ignored as mere beautifiers or nostalgic aesthetes, and their actions increasingly echoed the potential power and disruption of protest movements multiplying in the city at the time.

In this atmosphere, citizen-critics broadened their complaints. Repeatedly they described the market as a living organism and an especially potent symbol of a diverse landscape that had been or soon could be lost in the new consumer landscape downtown developers had planned. Like Steinbrueck, letter writers described the market's "special atmosphere" and the "real human element" that could "only be destroyed in the process of urban improvements."[93] The racial and ethnic diversity of this landscape, although always an element of Tobey's and the Friends of the Market's imagery, began to rise to the top of citizen-activists' concerns, at least rhetorically, reflecting the broader racial politics in the city.

Only a year before the march on city hall in 1969, after decades of pressure from civil rights groups in Seattle, the city council finally passed its first open-housing ordinance outlawing restrictive covenants and discrimination in housing. The same year, Seattle had become the first U.S. city to receive operational grants to help finance the construction of neighborhood centers, recreation centers, and housing in a Model Cities neighborhood. The *New York Times* described the area, Seattle's Central District: "Eighty-five per cent of Seattle's Negroes live in the neighborhood, making up more than half its population. Japanese, Chinese and a few American Indians, Eskimos and Filipinos also live there. Ten percent of the residents are white."[94] Such recent triumphs, as well as recent urban unrest, preoccupied the comments of petitioners, who wrote that the market was a place "where people of all races, ages, and ways of life can and do mix in harmony."[95] The market, these petitioners argued, held symbolic weight as the heart of the city and a potential "meeting ground" of the "poor," the "old," and "minority races."[96] Tobey's vision of the market as a diverse and democratic space therefore took on more significance in the heightened atmosphere of Seattle and national politics at the end of the 1960s.

But most petitioners linked their feelings about the market to their desires for a particular consumer experience that contrasted sharply with the CA's modernized downtown and their visions of white, middle-class female consumers. In this emerging discourse, the supermarket came to epitomize the lack of human and environmental diversity, the antimarket that constituted the focus of growing concerns about the sterile and heartless metropolis. A family that described purchasing all their fresh food at the market for over five years argued that the proposed redevelopment plan would replace it with "a structure which *in essence* will be not much different from a supermarket."[97] Indeed, in 1968 and 1969 the supermarket as symbol of a transformed urban and hinterland environment entered many letters. Melba Windoffer, for instance, argued that the market should remain unchanged because it was "such a welcome contrast to all these super markets."[98] Marilyn Skeels told the city council that the plan was "horrible," and remarked sarcastically that redevelopment would make the market "like one of those 'beautiful' A&P stores."[99] A woman from the Kent Valley described the proposal as "the slaughter of the Pike Place Market" and celebrated human diversity as well as the marked difference "from shopping in a supermarket of plastic bags and cardboard boxes." She asked, "Are we Americans perhaps too interested in efficiency to notice life?"[100] Julie Martin, another citizen-petitioner, admitted that the market was "poor and dirty," but like many she personified the place: "Its heart is beating strong," she said, adding that she hoped developers would not be allowed to remake it in the style of the "vast shopping complexes" north and south of the city. And another writer emphasized the opportunity that the market provided for "urban and suburban Seattleites to obtain food in fresh and unprocessed richness."[101] Like Tobey, a growing list of petitioners had begun to link urban spaces and their consumption with the health of a diverse democratic urban environment. They believed the future health of the body politic depended on it.

The March and April hearings in 1969 regarding the urban renewal scheme for market redevelopment saw large numbers of petitioners turn out. The presiding council member, Phyllis Lamphere, described the extended testimony: "We have been in session. We have met on 10 different days for 12 sessions and heard 33½ hours of discussion." Indeed the hearings created 866 pages of testimony, eighty exhibits, and according to one observer, "countless wrangles."[102] Responding to the surge of citizen participation, the council even extended written testimony into August, during which the bulk of the comments were made in letter form. The more formal exhibits and statements to the city council were presented by groups such as the Friends of the Market or by experts in the field of urban design and planning.

In addition to criticizing the urban renewal plans as essentially corrupt and unwise, these professionals made the most cogent links between the heart of the city and its hinterland, providing a broader context to the consumer landscape that letter writers echoed. The Friends of the Market emphasized the market's role as the antisupermarket, arguing that the market was a place "where young people who have never known anything other than precut meat, frozen vegetables or homogenized milk can discover some things they won't see on television or in Disney's picture books or movies."[103] But the most elaborate statement that linked what petitioners were saying to a broader theme of interconnection of the urban and hinterland landscapes through consumption came from the UW urban planning professor Ernst Gayden. His statement included a diagram of concentric circles around the market to show connections between the immediate neighborhood and the larger "metropolitan area." Using this illustration of "money" and "goods" moving in opposite directions, Gayden mapped what he called "the ecology of the market." Confirming the emerging holistic view expressed in the variety of letters and petitions, Gayden emphasized that the city's plan to preserve an acre and a half—rather than the seven acres that Steinbrueck and the Friends of the Market proposed for a Historic Preservation District—missed the point that the market was part of a "specific environmental context." Relationships, he said, "are essential to the health, even the existence, of the 'Market' as we know it, and . . . a plan which would eliminate the network of integral relationships and replace them with a new set of relationships which might relate to the 'Market' only as if it were merely any other shopping center, will in all probability so change the 'Market' that we might as well not bother to save it."[104]

54

By the end of the 1960s, the argument for the market's preservation had evolved into a remarkably intricate idea of an organic neighborhood. As the opposition to the CA's plans evolved, Seattleites watched different interests in the city debate the connections between the way they consumed and the kind of environment they hoped to support, inside and outside the city. And like Tobey's paintings, this vision of the sustainable city toward which they groped began to incorporate a discussion of inequality and diversity. With the market as a focus, these different social and political debates of the 1960s took on a spatial component, a physical form.

From Boom to Bust

Seattle reached a turning point in 1969. That year, the *New York Times* described a city experiencing growing pains: "Seattle today has nationally known restaurants, stores chic enough to attract Eastern fashion editors, a flourishing antiwar movement, a new cylindrical hotel, an important boutique on the wharf, the beginning of a hippie problem, and a fight over urban renewal." Depicting the city's economy as "still tied heavily to Boeing," the *Times* described a city that had grown fast and now faced "overflowing" freeways, the "brown haze" of pollution, and "racial tensions" that had "already produced several minor disturbances." By 1969 the sense of impending problems had built to a climax against the backdrop of a national recession.[105]

Beginning in 1969 the tide began to turn against the large-scale, expensive, government-financed plans that the CA and the city had envisioned. One reason was that money for such city-building projects had simply begun to run out. The confluence of various events shaped the market dispute's outcome specifically, yet it is important for understanding all the diverse urban environmental interventions that increasingly affected urban politics after the 1960s as well. The market was not only an obvious focus for historic preservationists but also a touchstone for the broader nexus of other political causes that challenged the status quo of urban politics after 1969. As the process of deindustrialization increasingly reshaped northwestern landscapes and politics, these changes gave new weight and persuasive power to the organic and holistic claims that Steinbrueck and Tobey had begun to articulate at the start of the decade.

Seattle felt this downturn acutely beginning as early as 1968. Boeing, tethered so tightly to federal government contracts and the health of the national wartime economy, helped determine the local crisis. Beginning in 1968, the engine of economic health in the region began to announce job layoffs, initiating a period known locally as the "Boeing bust." By 1969, after consecutive years of rapid growth in Seattle's postwar economy—including housing and small industry associated with airplane manufacturing—the aeronautics company began to lay off workers. In 1969 alone the company cut over 16,300 jobs. By 1970 the once robust local economy that had helped raise the spirits of local boosters and downtown builders was officially in a tailspin. This local downturn intensified the effects of a national recession and brought into stark relief existing problems with racial and economic inequality in the city. The urban uprisings in cities across the country, including riots in the Central District in reaction to Martin Luther King's assassination in 1968, only magnified these conditions. The petitioners' arguments gathered weight in this volatile atmosphere—but the downtown developers would make their rhetorical claims in the same context.

Local faith in the one-company-town economy had been shaken, and the usually ebullient local business press was in a defensive mood. In response to a general public grown suspicious of downtown businesspeople, *Seattle Business*, a newsletter published monthly by the Seattle Chamber of Commerce, ran numerous articles after 1968 describing how they were part of the solution while admitting to serious problems in the local economy. Responding implicitly to the events during the summer of 1968, the newsletter's editors wrote: "As the crisis in other cities has deepened, the response of the business community across the nation has been dramatic." The editor of *Seattle Business* described new alliances between government agencies such as the recently unveiled Equal Employment Opportunity Plan (EEOP, a division of the Chamber of Commerce) and local business groups in Seattle to "meet the challenge of finding jobs for the disadvantaged citizens and poor youth." The pages of *Seattle Business* in the late 1960s had already reflected a half-hearted commitment to creating more opportunities for people of color. But in reality the business community—and especially groups such as the CA—had focused less on inequality and much more on planning that involved moving federal funds into remodeling the downtown and capturing prized consumers. The events of 1968 and the beginning of the Boeing bust offered a chastening moment. The Chamber of Commerce described summer jobs programs and other efforts to expose "poor youngsters from the Central Area" to the "ways of life in business." After 1968 the business community was finally ready to "bring those who have felt alienated into the mainstream of our economic society."[106]

Seattle Business understood as did few others in the community what the downturn would mean for the local economy, and it was also aware of the business community's own compromised position. In retrospective articles about the decade's successes, the newsletter seemed nostalgic for a bygone age, a postwar era that saw "surges of industrial, commercial and residential growth." But the 1969 downturn suggested a worrisome slide backward into the area's historical economy of resource extraction. The decline of Boeing, the director of research for the Seattle Area Industrial Council (SAIC) explained, "involved more workers than [were] employed in either of the region's next two largest manufacturing industries, lumber and food processing." In 1969 he warned that the Seattle-Tacoma-Everett area was "in for a period of economic contraction throughout most of 1970." Reiterating the SAIC's concern for youth employment, the director wrote that 1970 would be the "first stern test of the effectiveness of community efforts to train and place minority and hard core unemployed on a long term basis."[107]

With the economic picture worsening in 1971, the common sentiment of Seattleites was evident in a billboard along the highway south of the city that read, "Will the last person leaving Seattle—turn out the lights." As workers left Seattle for better prospects down the road, the Chamber of Commerce admitted that the problem was "unusually severe in the area" but portrayed the "major cause of unemployment" as being "layoffs in a single industry brought on by conditions in the national economy." Norton Clapp, the Chamber of Commerce president, noted that "not all the lights are out in Seattle—they are burning long and bright in the offices of business and government leaders who have confidence in the future promise of our city."[108] But the kinds of promises and plans that the CA and the still-compliant city government had made in the past were waning at the very moment when an activated citizenry began to take hold of the reins of urban environmental problems. Downtown businesspeople's arguments for using quickly dwindling public money to bulldoze the market began to ring hollow to a public that appeared increasingly mobilized and critical of the status quo. And in their proposed plans each side in the contest invoked the city's economic viability within the new constraints of a poor economy and social unrest.

Writing to Steinbrueck early in 1969, when he and his citizens' movement launched their large-scale challenge to the redevelopment plan, Mayor J. D. Braman expressed the economic revitalization argument that the CA had long endorsed. Referring to the market as a "nostalgic portion of the city," the mayor wrote that it "must be developed in an effort to maintain the health of the downtown district as an area in which we hope to be able to regain as residents many of the people of high income who have fled outside of the city, leaving the center core to deteriorate."[109] The mayor reiterated the persistent logic of consumer flight while revealing the class-skewed goals for urban renewal that appeared increasingly shortsighted and bankrupt in the quickly shifting political winds of the time. Still heeding this logic in August 1969, and despite the building protest movement, the city council rubber stamped the urban renewal plans for the market.

Steinbrueck fought on, asking in a September press release for "the abandonment of the pretense of Urban Renewal"; he argued that the "program was never intended for this type of colossal, block-buster, business district glorification." The "citizenry is aroused and their often expressed concern for this unique part of the city's life must be recognized," he wrote. In parentheses he added, "or is only violence to be heeded by city hall these days?" Steinbrueck effectively invoked the broader context of urban unrest and protest to make his point and pressure the city. Meanwhile, beginning in 1970, the Friends of the Market proposed a ballot initiative for the following year's election that would ask citizens to approve the preservation of seven acres of the original market site as a Historic Preservation District.

While they collected signatures to put the initiative on the 1971 ballot, the Friends of the Market simultaneously turned to the U.S. Department of Interior in an effort to block the development on the grounds that the market was a historic neighborhood worthy of preservation. The Central Association could barely disguise its outraged reaction to this effort in its summer newsletter in 1970. After more than ten years of sustained effort, the central feature of its plan—urban renewal monies, passed by the city in 1969—appeared in jeopardy because of these activists. The CA seemed especially angry at the guerilla tactics of the activists, who had sought the

Benefit rock concert at Seward Park for Pike Place Market, 1971. *Seattle Post-Intelligencer* Collection, image no. 1986.5.54480.1. Courtesy Museum of History and Industry.

federal designation "without consulting" the mayor "or any City agency responsible for the redevelopment project which had been approved last year by the City Council after lengthy hearings."[110] The CA warned that historic preservation would block federal approval of the $9.8 million that would allow the city to "move ahead with the long-planned project," which it argued was the market's only possibility for survival. If the plan was "put on ice" the CA warned, the city would "have to enforce building and housing codes in the area," which would "mean demolition of most buildings" anyway.[111]

In a 1970 press release the Friends of the Market pointed to the Northwest's economic woes to make their argument as well. Not only would the city's urban renewal plans fail to aid historic preservation of the market, the group maintained, but it would fail to aid the economy of Seattle, too.

The city's plan, in addition to replacing the existing market with something "new and sterile," would not help "the ailing Seattle economy." The Friends of the Market argued, "While our depressed economy increases the need for the people's low cost market, it will be gradually eliminated and more people put out of work." Reflecting the growing criticism of urban renewal, they added that using Housing and Urban Development (HUD) money in this way would only be "filling the pockets of property owners" while "putting the famous federal bulldozer to work in demolition." The Friends of the Market argued that urban renewal funds would stimulate the economy only if they were used differently, to truly rehabilitate the neighborhood: "People could be employed almost immediately for rehabilitation of buildings in the true spirit and purpose of urban renewal."[112]

In the midst of the local Boeing bust and the nationwide economic downturn, the links among consumption, democracy, and environment were further underlined. The early outlines of a new consumer sensibility were evident in Tobey's paintings and in the antimodernist and antisupermarket rhetoric of the early 1960s protest movement, just as they were evident in the CA's desperate attempts to draw consumers to the core by any means necessary. But neither side at the time fully understood how this contested consumer sensibility would take shape. Both sides in the debate over the market made implied and overt arguments about the role of consumption in the regional economy. Yet downtown businesspeople and the CA still tended to see the economic crisis of the early 1970s through the lens of industrial production, and for good reason. "There is no question," wrote the president of the Chamber of Commerce in 1971, "that our area is in a serious condition as for its levels of unemployment and economic curtailment." But he emphasized external factors including "temporary stoppage on projects such as the development of the Alaska North Slope oil fields, and construction of the Trans Alaska Pipeline System." The Chamber of Commerce focused on the nuts and bolts of the economy's downturn while predicting "economic stabilization in the near future."[113] In comparison, the Friends of the Market's ideas about the health of the local economy appeared ill-informed and short-sighted to the CA, whose members felt they operated on a much larger scale. But both sides fought for the hearts of voting consumers in the struggle over urban space.

David Brewster, a longtime Seattle journalist and observer of city government writing in *Seattle Magazine* during 1970, described the shifting economic winds of the time, predicting the decline of the existing power bloc. The CA was a "classic illustration of what is sometimes called government by economic notables." Drawing on a UW sociologist's examination of Seattle, Brewster listed the powers in order of significance: "1. Manufacturing Executive; 2. Wholesale owner and investor; 3. Mercantile executive; 4. Real estate executive." Yet this power bloc was far from unified, according to Brewster, who saw "countervailing forces" at work locally by the late 1960s. With the rise of "competing organization like community councils" and "government itself," cracks seemed to be appearing in the monolithic downtown power structure.[114] Indeed, Steinbrueck's group had the support of a rising class of urban actors in the 1960s, many of whom represented the design professions, the university, and a new class of younger professionals. In 1970 these "countervailing" forces elected a young, progressive mayor and a city council that was committed to reform and that championed the 1971 ballot measure to preserve the market as a seven-acre historic site.[115] This shift in urban sensibility—from a producer to consumer economy and toward an economy more centered on services and tourism—was only forming in the late 1960s and early 1970s. Steinbrueck and the Friends of the Market were no more savvy than these downtown businesspeople, though they at least rhetorically challenged the CA's consumer-driven vision. But as the historian Judy Morley has argued, their plans for the market were no less consumer driven than the CA's, despite their emphasis on the market's authenticity, diversity, and links to the more organic rhythms of the past. The question became, What kind of consumption patterns would downtown Seattle adopt—with the intervention of the federal government in the form of funds for Historic Districts or urban renewal—as part of this new urban sensibility?

The answer to this question reveals the link between consumption and the emerging urban environment, and between consumer desire and the postwar environmentalism. In a 1971 article titled "Keeping the There There," the *New York Times* architecture critic Ada Louise Huxtable observed a trend she saw appearing throughout the country. Local officials from Tallahassee, Florida, to Pittsfield, Massachusetts, had begun linking environmental and historic preservation issues to larger economic growth strategies, especially with respect to consumption. "Ironically," she wrote, such preservation projects "are not the work of little old ladies playing house, but of unsentimental business interests."[116] The influential architecture critic, pointing out projects in San Francisco and Washington, D.C., that utilized "a similar formula for adapting fine old vernacular construction of both architectural value and historical recall for workable contemporary purposes," noted that in contrast, Seattle seemed "determined to get rid of its waterfront farmers market" at the very moment when many cities had begun to "turn back their waterfronts to public uses and pleasures." She asked incredulously, "What greater pleasure than fresh salmon and strawberries with a view of the Sound?" Six months before the initiative vote on preserving the market, Huxtable predicted its survival. "Too many have learned too much from watching the wreckers, and results."[117] The CA had not yet caught on to this strategy; indeed, it had fought for the market's real estate for so long that it seemed dead set to modernize the site even as such plans began to appear less fashionable.[118]

At issue, though, was not merely the spaces or buildings but also what was inside the market. Whether the Pike Place Market was authentic, worth preserving for architectural reasons, or a unique neighborhood worth restoring, it was the antisupermarket aesthetic and consumer pleasures associated with the place that made it part of a larger consumer shift by the 1960s and an early emblem of the sustainable city. Soon after Huxtable linked the market to the joys of salmon and strawberries on the shores of Puget Sound, *Gourmet* published a similar feature on it that helped to further explain the proto–slow food movement as part of the urban aesthetic that the Friends of the Market hoped to create. In the article, city, hinterland, and consumption merge in a seamless postindustrial consumer landscape that the CA seemed unable to grasp in this rapidly shifting atmosphere of tastes and politics. Describing Seattle and the market in terms associated with endangered species threatened by creeping modernity, the writer for *Gourmet* remarked, "I was glad to have an opportunity to see it while it lasts." The article detailed for its national audience the fight over the market while helping to reiterate and cement an image of Seattle's mix of nature and urbanity that would soon become iconic: "King salmon, a best buy at eighty-five cents a pound. The noble salmon, whose indomitable instinct to return to the inland rivulet where it was spawned is one of nature's most awesome mysteries, comes to the three fish stands at the Market only hours from the sea."[119] *Gourmet* illustrated this crucial set of connections among consumption, nature, and urbanism that would be crucial to the market's image and part of Seattle's postindustrial reinvention.

Despite their standoff with downtown developers, the market's supporters held the secret to the city's future success in exploiting the city's connections, real and imagined, to nature. *Gourmet* thus echoed the growing sentiment of local citizens who had decried the supermarket fate of a remodeled market—and city—in the early 1960s.

In the months leading up to the 1971 election that included the ballot measure, the CA and its allies tried unsuccessfully to muddy the waters of debate in the city.[120] The *Seattle Times* in particular did little to clarify the competing positions of the city and groups such as the Friends of the Market. It did not help that by 1971 the latter had an associated group called the Alliance for a Living Market and the City, and the CA had invented another group called the "Committee to Save the Market." These groups lined up on each side of the essential issue, but the number of "grassroots" groups added to voters' confusion, which the CA and its allies exploited. Essentially, Seattleites had to choose one of two plans. The city and CA wanted to refurbish only 1.7 acres of the original site and designate that area as a unique and historic, while the rest of the site, according to the *Post-Intelligencer,* "would have its old buildings leveled." The property would then "be sold as sites for high-rise view apartment houses," with some sections preserved for "open space."[121] This development would garner up to $10.6 million in federal urban renewal funds. In clear contrast, the Friends of the Market and the Alliance for a Living Market hoped to preserve over 7 acres of the original site and create a historical commission to manage and rehabilitate the neighborhood. Despite the numerous players in the battle, it all came down to these two ideas, two sets of funding, and two notions of the market's value to the city.

Seattleites went to the polls on November 2, 1971, and approved the preservation ordinance. The proclamation, signed on December 1, 1971, certified the initiative. The new and youngest mayor in the city's history announced "an Ordinance to preserve, improve, and restore the Pike Place Markets, creating the Pike Place Market Historical District, prohibiting alteration, demolition, construction, reconstruction, restoration or modification of structures therein without a certificate of approval, establishing a Historical Preservation Commission, and providing for administration and enforcement."[122] The details of this new designation would open another innovative chapter in historic preservation as the city and the new commission debated and decided the future of the market in the following years. But the market debate showed more than anything else a clear shift of power in the city and a shift in urban sensibilities. As if to punctuate these massive changes, Ben Ehrlichman, the developer and former Central Association leader who had been so committed to the redevelopment plan, died just two weeks before the vote, on October 24, 1971. The electorate's decision that would finally block his plans came just a week after his funeral.

By the time a critical mass of Seattleites had joined the fight to save the market by the late 1960s, the flush postwar years of urban renewal and federal largesse were already coming to a close, and a new atmosphere of urban politics framed the rhetorical debate about the city and the environment to follow. Few at the time understood or could predict the stingier urban policies or the economic, political, and environmental "crises" that the city and the country would face during the 1970s. But the market debate heralded this emerging context and suggested the basic contours of a broadly defined urban environmental activism. Grassroots energies that focused on the market and then later on more obvious environmental concerns in Seattle had a lot to do with the way people consumed and the consequences their consumption had for the places they loved. Tobey's "processed cheese" was an apt metaphor after all. At the market people expressed the beginnings of a postwar environmentalism that began to link elements as seemingly disparate as city and hinterland, food and politics, personal identity and cheese.

2

Neighborhood

While developers and their critics battled over the future of the Pike Place Market in 1969, less than three miles away, in Seattle's Central District, an integrated team of neighborhood organizers set out on their first annual "Fall Drive on Rats." The group, including white college students, African American residents, and Asian American activists from the city's International District, reportedly inspected 10,090 manholes and used over 500 pounds of poison bait in their effort to rid the area of rodents. The city government had neglected the problem for years, but that fall citizen-activists reinstated a "rodent proofing" program in the Central District as part of a new plan to improve health and other environmental conditions.[1] By 1970 the volunteer rat counters had created a baseline survey of the neighborhood, examining over 12,000 homes and vacant lots and "measuring 57 variables for each property."[2] By the end of the 1960s, the sight of activists walking down the neighborhood's alleys with clipboards in hand seemed normal. Central District residents watched from their windows as newly trained "environmental aides" combed 661 city blocks and 104.2 miles of alleys to assess everything from "refuse storage, rodents," and other "vectors" to "exterior building conditions, safety hazards, air pollution, noise, odors, alleys, sidewalks, natural deficiencies and abandoned houses."[3] In addition to holding press conferences at the time and inviting some local news coverage of their efforts, the group erected imposing billboards

Billboard advertising Model Cities Rat Project, July 1, 1970–Sept. 30, 1970. Model Cities Program Environmental Health Project Records, Photograph Collection, item no. 69314. Courtesy Seattle Municipal Archives.

in the neighborhood depicting the rat problem. "Can Rats Get Into Your House?" one of the billboards asked.[4] In bold block letters at the bottom of the attention-grabbing image appeared a phone number for the ubiquitous Model Cities (MC) Environmental Health Project.

In the following years, as part of the comprehensive Model Cities program in Seattle, the Environmental Health Project, along with other efforts to improve the physical environment of the neighborhood, did far more than count rats. The Model Cities program (originally the Demonstration Cities program) was President Lyndon Johnson's last-gasp attempt to expand and reinvigorate the War on Poverty against the backdrop of riots, stark failures of urban renewal, and a growing public perception of an urban crisis in American cities.[5] Model Cities' basic goals were to better coordinate federal, state, and local resources to fight the roots of poverty; to develop innovative and responsive local programs; and above all, to require the direct involvement of local residents in the planning process.[6] The short-lived program, lasting from 1967 to 1974, when President Richard Nixon canceled it, left a profound and lasting influence in Seattle's neighborhoods, for it gave them the power and tools to define locally relevant and successful projects at a time when the U.S. government's credibility was in question among conservatives in Congress, many urban activists, and especially African Americans.

People living in Seattle's Central District had particular reasons to mistrust federal and local officials in the early 1960s. City planners and the mayor expended tremendous political energy and money on grandiose downtown redevelopment plans, such as the proposed redevelopment of the Pike Place Market. They emphasized modernization to create a consumer utopia for white, middle-class shoppers while actively beating back civil rights activists' efforts to pass open-housing initiatives in the city. Although the Central District never matched the profile of other large ghettoes in the United States, because of its higher rates of homeownership and single-family dwellings, the area still suffered from the worst problems associated with segregation and the negative consequences of metropolitanization, which redirected money toward urban renewal and suburban development and away from the city's older neighborhoods. One contemporary assessment noted that there was "far more deterioration, poverty, and deprivation in the Central Area" than first met the eye and that conditions had been "growing worse."[7] As the overall population of Seattle and King County grew between 1940 and the 1960s, high numbers of African Americans joined the growing population moving to the urban West in search of higher-paying jobs.[8] In Seattle, because of segregated housing markets, this population growth focused on the Central District, which became increasingly densely populated. Many homes in the area had been chopped up into multifamily dwellings during this period, and according to the 1970 assessment, many of the houses were far below the city's or nation's standards.[9]

Cramped conditions and concentrated poverty exacerbated the long-term ill effects of racial segregation. One third of all Seattle's welfare recipients, for instance, lived in this 2.25-square-mile area due west of the city's oldest part. Residents there earned 27 percent less in median annual income than did those in the rest of Seattle. The infant mortality rate was three times as high as that for the rest of the city, as was the unemployment rate. An estimated 18 percent of the housing in the area could be defined as dilapidated, whereas only 12 percent fell in that category in the rest of the city, and 32 percent of the housing required major repairs. Finally, 85 percent of the population in the historically black Model Cities neighborhood was African American, fully half the entire city's African American population. Essentially, Seattle's ethnic minorities and urban poor were concentrated in the Central District. According to a Model Cities report at the time, "the centers of Seattle's Chinese and Japanese communities" were "also within the MN [model neighborhood]. Other minorities include American Indian (4 percent); Eskimos (3 percent); Filipinos (3 percent); and Caucasians (10 percent)."[10] The MC's own report suggested the diverse population in the area but was less clear about the historical importance of the neighborhood as a cultural crossroads.

Despite its problems, Seattle's Central District had developed a strong sense of place and collective history over several decades, especially after World War II. Indeed, like the Pike Place Market, which found a champion in Mark Tobey and his famous and inspiring images, the Central Area had its own artistic chronicler in Jacob Lawrence, who moved to Seattle during the MC period and made some of his most important paintings depicting African American communities and life there.[11] The life he portrayed, however, was complex, for as was true of other racially segregated neighborhoods in the United States, segregation in the Central District could be both stifling and empowering.

The identity of a group of people cannot be reduced to oppression. The historian Quintard Taylor describes the problem of defining a community only in such terms. The African American community "has been defined by denial and exclusion," he writes; nonetheless, although such factors have been important "in shaping black urban life, . . . they have not exclusively determined the nature" of that community." He argues that ultimately, people come to define their sense of collective identity and values through an infinite variety of interactions. "Thus it is crucial for the urban historian," he writes, "to examine how institutions and organizations service that process and forge it into a black urban ethos."[12] In Seattle these

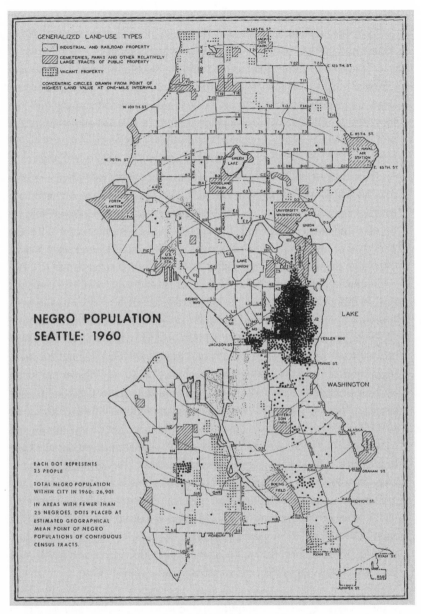

A map of Seattle showing segregation of African Americans in 1960. From Calvin A. Schmid, *Nonwhite Races of Washington* (Olympia: Washington State Planning and Community Affairs Agency, 1968).

69

shaping forces came from within and without, as the history of the Model Cities program there confirms. Moreover, African Americans and Asian Americans there forged their identities in close proximity to each other in a multiethnic neighborhood, often defining a common sense of collective political purpose. In Seattle, then, the federal Model Cities program would build on this complex and collective local neighborhood identity and history. The program significantly helped reinforce this sense of community and sense of place among people in an already diverse and active neighborhood.[13]

This sense of place helped to heighten organizers' sensitivity to conditions that straddled the line between the social and the environmental. Segregation often manifested in environmental consequences that inhered in physical space, both the built and the natural worlds. Like the preservationists who fought to save the Pike Place Market as a living, breathing neighborhood, organizers in the Central District began to grapple with their own environmental, planning, and even aesthetic concerns during this period. They also sought to solve on their own terms the environmental problems that affected their communities. Unlike the market, however, the Central District was far from a cause célèbre. Few in Seattle's power structure or arts community had nostalgic feelings about the area, even though it was the indisputable center of Seattle's African American and Asian American communities. Yet the work of Model Cities volunteers and mainstream grassroots activists shared common elements with the efforts of those who fought for the market. Although their campaign differed in tone and outcome, the Central District activists similarly organized their own neighborhoods and sought to control the production of space in the city. They too hoped to stimulate civic participation and assert local autonomy; armed with a strong sense of place, they saw their neighborhood as the heart of the city. And in racially divided Seattle, activists found that their environmental problems were obviously entangled with racism and poverty, a connection they felt viscerally each day in the segregated and even physically dangerous environments where they lived.

Yet environmental historians and historians of the 1960s have done little to examine the wide-ranging influence of Model Cities programs in contributing to this kind of urban environmental understanding. The still-dominant narrative of environmental activism focuses on the activities of groups such as the Sierra Club or Audubon Society, privileging certain ways of knowing and thinking about nature over others. These stories rarely feature cities and urbanites—both people of color and integrated groups—as pioneers or innovative activists in the environmental movement. Similarly, the origins of the environmental justice movement and urban activism are usually plotted much later, in the 1980s.[14] Historiography of the 1960s has also tended to separate the history of environmentalism from the discussion of urban politics. In fact, however, deep in the "urban crisis" of the 1960s, and just as the popular environmental movement began to emerge, concerns about the environment were no less a concern in the Central District than they were in the more affluent or predominantly white parts of the city. But in the Central District nature could seem dangerous, a threat to the physical bodies, homes, and futures of people.[15] This sense of danger could be mobilized to achieve a variety of social goals as well. Assessing the environmental aspects of Seattle's Model Cities program, therefore, suggests important roots of sustainability, environmental justice, and environmental activism in this episode in the War on Poverty. Grassroots experimentalism in Seattle and elsewhere reinforced links between social and physical environments and between everyday physical bodies and nature. Like their counterparts who hoped to preserve a particular urban ecology at the Pike Place Market, activists in the Central District consciously linked their politics with places. The struggle for civil rights and the War on Poverty are rarely portrayed as coevolving with the postwar environmental movements,[16] yet they are entwined in important ways, and their confluence would have a lasting influence on the politics and outcomes of grassroots organizing, including the kind of city this organizing has left in its wake.

The Federal Model Cities Program in Seattle

Postwar disinvestment in cities and subsidies to suburban development contributed to deterioration in housing and infrastructure that could create serious environmental problems in inner cities, just as these same forces could help subsidize and expand the metropolitan environment. As MC Environmental Health Project activists surveyed their neighborhood for rats, inspected dilapidated housing, or began to address the dearth of green space in their community, they consciously connected these environmental and human problems to blocked access to capital, jobs, and political power. The environmental initiatives of the Model Cities program—

such as the Fall Drive on Rats—provided tools, education, and in many cases, jobs. The program helped to inspire a variety of actors to begin refining their own neighborhood-level utopias. The Central District and other MC neighborhoods each had deep roots in the city's collective history. But the introduction of the Model Cities program, like the market controversy, began to draw residents' and activists' attention to environmental problems at a crucial moment in Seattle's and the nation's history.

Urban planners, preservationists, and civil rights activists nationwide shared a common distress about the urban environment in the 1960s, even though their solutions and emphases were often at odds. Just as the Pike Place Market controversy was a distinctly local conflict that drew from local sentiments about history, sense of place, and environment, the specific environmental concerns of Model Cities activists in Seattle, too, would emerge from a local context and set of priorities. Yet neither of these neighborhood-focused movements was wholly local. National events, national discussions, and especially federal policies such as urban renewal, historic preservation laws, and the War on Poverty helped to frame local possibilities. This dynamic is important because it suggests how environmental politics—and especially the urban-based counterculture green movement that would follow in Seattle and other cities on the West Coast—were not revolutionary expressions of do-it-yourself pioneers or small is beautiful advocates. Such local efforts could not have happened without federal support, and the work of neighborhood activists in the following decade owed much to the example of civil rights and community organizers well before the popular ecology movement. In Seattle, however, federal policy and local initiative would combine with unintended consequences. In the case of Model Cities, it is worth understanding the framework that policy makers put in place at the federal level and its implementation in everyday practice by the activists who shaped its environmental character. The seeds of a unique urban environmentalism took root in this dynamic interrelationship of local civil rights culture, specific neighborhoods, and federal policy.[17]

In his speech introducing the Demonstration Cities bill to Congress in 1966, Lyndon Johnson painted a picture of the new metropolitan landscape of the United States that many Seattleites could understand: "If we devour our countryside as though it were limitless, while our ruins—millions of tenement apartments and dilapidated houses—go unredeemed . . . [;] if we become two people—the suburban affluent and the urban poor, each filled with mistrust and fear one for the other . . . [;] if this is our desire and policy as a people, then we shall effectively cripple each generation to come." Johnson's groundbreaking bill called for an expansion of the War on Poverty to build "an effort larger in scope, more comprehensive, more concentrated, than any that has gone before."[18] His speech was not mere rhetoric. The president outlined a more decentralized government program that shifted emphasis toward organizing activists around a plethora of local initiatives related to their local environments.

Before 1966, urban renewal plans, such as the Pike Place development plan, had emphasized commercial real estate rehabilitation. These efforts often shored up the fortunes of downtown developers at the expense of low-income housing, while often actively excluding African Americans and poor people from the planning process and its potential benefits. The federal government handled urban problems related to health and welfare separately from physical infrastructures, although demolition and removal of entire neighborhoods was often rationalized as an antipoverty measure. In 1966, the enduring problems of race and poverty in the United States were still primarily understood and addressed in terms of blight. But blight is an imprecise concept.

In his study of postwar Oakland, California, for example, the historian Robert O. Self argues that most powerful white businesspeople and officials in this midsized city saw its physical and economic problems as products of blight rather than symptoms of "social inequality." Local white decision makers with this point of view, he argues, believed that blight originated not "in the racial segregation of housing and labor markets or in the unequal distribution of political and economic power, but in the deterioration of aging housing stock, overcrowding, and declining property values."[19] The history of urban renewal in the United States certainly demonstrates this willful disregard for structural racism and poverty. Seattle's local planners and developers shared it, boosting proposals such as Pike Place Market renewal well into the 1970s while simultaneously blocking open-housing and fair employment laws in the city. And the Model Cities program itself, because it relied on neighborhood initiative and the coordination of local government, could often be hijacked by mayors and interests who had no desire to help empower neighborhoods or civil rights activists.[20]

At first glance, then, the Model Cities program could appear to have perpetuated the core misconceptions about urban poverty that Self underlines in his study of Oakland and that many urban activists could clearly understand. However, it departed from this prevailing approach to urban renewal and poverty significantly. Unlike all the program's predecessors, MC plans explicitly linked rehabilitation of the urban environment with social factors including access to employment, desegregation, and poverty. Johnson's Model Cities task force crafted a plan that would include "massive additions" to low-cost housing in something its authors called "a total approach" to "combining physical rebuilding with human concerns." At its core, Model Cities, unlike earlier poverty programs, was committed to a broader "social program," or what Johnson referred to as the "total welfare of the person," as well as the improvement of the "physical city in which he lives." To avoid the stark failures of urban renewal, Johnson's Model Cities program urged cities to "maintain or establish a residential character in the area," suggesting a new sensitivity to the social and physical fabric of existing communities. The new program also offered an unprecedented effort to coordinate resources and mobilize local groups to ensure that, as the MC task force would agree, "the key decisions as to the future of American cities are made by the citizens who live there."[21] Significantly, the program's "increased flexibility"[22] created an innovative framework within which Seattleites began to ever more clearly construe for themselves the links among poverty, racism, and environmental problems. In addition, Johnson expressed his commitment to desegregation in housing at a time when many cities had yet to pass open-housing legislation. At "the center of the cities' housing problem lies racial desegregation," he explained.[23] In 1966, he pushed not just for Model Cities but also for legislation that would bar discrimination in the sale or rental of housing.

Johnson's broad experiment increased federal influence and funding for cities while decentralizing power and decision making by placing them in the hands of local government and especially activists and citizen groups in depressed places such as the Central District. But unlike Community Action and other Office of Economic Opportunity programs that empowered citizens in the early 1960s and helped to build Seattle's civil rights infrastructure, the MC brought the social and the environmental together in new and productive ways, allowing local groups more power to define changes in their environments. Along with other neighborhood-based initiatives and existing War on Poverty programs in the Central District during the 1960s, Model Cities helped to reinforce and even elaborate a building movement for neighborhood power. This neighborhood-level organizing marked a clear contrast to a decade of downtown dominance in city government and planning, part of a larger shift that had begun as early as 1960.[24]

Seattle's existing civil rights movement was highly organized, with decades of local organizing experience, helping Seattle to have its Model Cities development plan approved in 1968, before any other city.[25] Planning and implementing changes in the community revealed this dynamic political context already in place in the neighborhood. At the same time, clear divisions existed within the community by the late 1960s. Black Power advocates at odds with the civil rights establishment were suspicious of the program. Representing the shifting political ground of the time, some advocates in Seattle voiced such a critique of MC program, fearing the status quo of earlier programs that emphasized environmental solutions to urban ills while neglecting jobs and the desegregation of housing markets. This local division in the community was apparent in the tension between the younger activists, who preferred a more confrontational approach, and the older civil rights establishment. In particular, Seattle had strong "Red Power," "Yellow Power," and Black Panther Party activists by the late 1960s. These activists would form a crucial element of neighborhood organizing during the 1970s. Ultimately, however, the tensions within the community contributed to an atmosphere of critical local oversight and transparency when neighborhood plans did finally emerge from the Model Cities experiment.

A year into the program, for instance, an editorial writer for the fairly radical *Afro American Journal* echoed the sharp critique of federal influence in the neighborhood, referring to the program as the "Model Cities Hoax": "It appears that when a people are in economic depression as black people are, you decorate their neighborhood."[26] The *Afro American Journal* conceded that the "primary work of remodeling a neighborhood is the work of builders and urban planners," but it added, "equally important will be the people's ability to earn sufficiently to maintain the improved property."[27] Speaking directly to the issue of jobs, an editorialist for the paper asked, "Where is the program that will produce 3000 Black men jobs in the skilled building trades industry that would be comparable to other communities of equal size?"[28] As historians of the civil rights movement in northern cities have demonstrated, fights over segregated work-places and property markets were always a local political issue, especially when the federal government failed to enforce the law. African Americans also famously organized and fought hard to desegregate the construction trades in Seattle throughout the 1960s. The *Afro American Journal* reflected this period's neighborhood-level struggles over approaches to civil rights. The critique astutely underlined liberalism's national failures while it revealed the simmering generation gap between the mainstream civil rights movement and the Black Power movement. Many in the neighborhood saw MC as yet another attempt to create more window dressing while mollifying residents during the long, hot summers of the late 1960s.[29]

Yet in Seattle the local, citizen-driven emphasis of MC seemed to hold up well in the process of the program's implementation.[30] The Central District was well prepared to use MC funds. For example, Charles Johnson, a prominent mainstream civil rights leader and local National Association for the Advancement of Colored People (NAACP) activist involved in Model Cities, recalled that the program "was a godsend for the civil rights community . . . because it gave an organization with money and with staff to help" those individuals "who had been volunteers doing this sort of thing out in the community."[31] Not only did Model Cities provide financial and planning support, but it also allowed citizens to determine—up to a point—how these funds should be used. With each initiative, according to Johnson, people in the community used MC money to desegregate municipal jobs with the police and fire departments, deploying the funds to train and prepare men and women for the service exams. "We had money to work with to do a lot of things to help integrate," he explained. As had few programs before, Model Cities offered more than just job training.[32]

By 1970 over 150 cities had begun the planning process to obtain federal Model Cities grants. But unlike their counterparts in other cities, Seattle's activists and mayor made environmental concerns a central feature of the eventual plan that citizens devised.[33] Seattle's Central Area Motivation Project (CAMP), an important local civil rights organization founded by neighborhood activists in 1964, led the charge, guiding local planning efforts before and after the grant was approved. Under the leadership of the civil rights veteran Walt Hundley, the widely respected CAMP would be crucial to the process of implementing Model Cities and exemplified the program's local drive. The *Seattle Trumpet*, CAMP's newspaper, described

Seattle's MC application as having been "drafted by a multi-racial group of 175 citizens" in the spring.[34] That fall, CAMP headquarters hosted a series of meetings that included church groups and other community organizers in an effort to define the neighborhood's future needs and to prepare for the impending Model Cities planning process. The neighborhood task forces that assembled, the *Trumpet* explained, would "underscore, and in many cases carry further many of the ideas set forth in the Model Cities proposal unveiled in April, 1967."[35]

The task forces highlighted community concerns about consumer protection, employment, housing, recreation, and law enforcement. But the most detailed section of the *Trumpet's* report on task force activity featured an extensive section on housing, recreation, and environment, revealing the program's mainly environmental trajectory. Displaying this early urban environmental approach, the findings of the task force reiterated the "crucial role" that "housing and healthy living space" played in moving African Americans "from the status of ghetto-dwellers to full citizens of the city." Homeownership formed a center of the task force recommendations, but the report also emphasized rehabilitation of existing housing. As perhaps its most forward-looking feature, the program created a plan for cooperative housing and local financing rather than public housing, for small landscaped parks and open space, and for a mixture of apartments and retail in new developments. Such ideas offered early expressions of some of the essential ideas that would animate "New Urbanism" and movements for sustainability in the following decades. Citizen involvement was critical to the entire process. "The physical form of our community," the report's authors concluded, "must be derived from the needs of the people within it. It can be seen that it would be disastrous to continue the trend of imposing the form from without, whether it be based upon calculations of highway engineers, the pressure of downtown expansion, or social prejudice." By November 1967, when Housing and Urban Development (HUD) officially announced that Seattle qualified to receive Model Cities funds, the neighborhood was well prepared for the next stage of planning, having already defined residents' prime areas of concern and commitment to transparency.[36]

Reflecting the strength of local civil rights organizations and CAMP's quality of leadership, the city appointed Hundley to be the director of Model Cities in Seattle. Hundley, a social worker and Yale Divinity School graduate from Philadelphia, would be a crucial link between the city and the neighborhood, helping to shepherd the planning process along through the municipal and federal bureaucracies; the *Seattle Times* recognized him as "one of several negroes who have moved into positions of leadership affecting the entire city."[37] Individuals such as Hundley epitomized the early civil rights commitment to desegregation and integration, unlike the younger activists in the neighborhood. Far from a radical, he would function as a familiar go-between with the city's powerful at a crucial and volatile time in Seattle. Hundley, remembered for his ability to defuse confrontations between the community and the mayor, would go on to be the city's director of management and budget during the 1970s. In 1977 he became superintendant of parks and recreation, overseeing a rapid expansion of the parks program. With Hundley's help, the previous season of organizing led seamlessly into the next step of the grant process, in which the community would meet to create committees, asses goals, and submit a comprehensive plan by October 1968. That year the program also chose its logo, a stylized image of a seagull and the catchphrase "Make It Fly." The shape represented the approximate boundaries of the 2½-acre Model Cities neighborhood that citizens and the city had first defined in their 1967 proposal. At the core of this region lies the Central District of Seattle, the epicenter of the War on Poverty in the Northwest and the home of CAMP. The historical center of black Seattle, then, formed the backbone of this new program.[38]

Despite this core constituency, the oddly shaped map expressed an encompassing physical and cultural geography and explained the project's broad appeal from the start. Although mapping the basic outlines of the most segregated neighborhoods in Seattle, the first Model Cities project there included a highly diverse set of neighborhoods and people within the larger area defined as central Seattle. To the west the map included the International District, dominated by Chinese, Japanese, and Filipino Americans. The western extent, the bird's feet, also encompassed Seattle's old Skid Road, or Pioneer Square, a recently designated historic district and home to many of the city's Native Americans. To the far east, at the head of the bird, the map included mixed neighborhoods of mainly whites and African Americans, including the Leschi and Madrona neighborhoods. The north section of the map reached into the southern flank of Capitol Hill, another racially integrated neighborhood at the time.

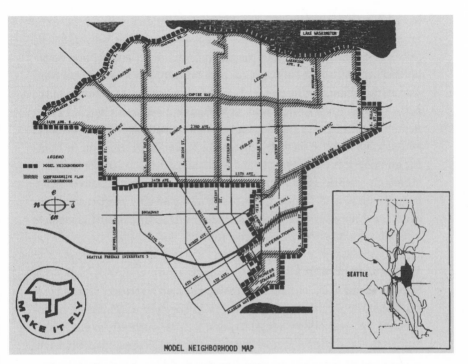

"Make It Fly" logo and map of the Seattle Model Cities program. From *Seattle Model City Program Third-Year Action Plan* (Seattle: Model City Program, 1971).

These borders in part explain why Seattle's Model Cities proposal succeeded, for the area was defined, according to one spokesperson, to be "as fluid as possible, so that it would encompass common problems and be workable for various programs without being the exclusive territory of the particular group."[39] Including Pioneer Square in the MC plan also expanded access to urban renewal grants as part of the project's mandate to coordinate with local funding and ongoing federal projects. The Model Cities neighborhood drew from a sense of place and history of multiethnic organizing. The encompassing and diverse constituency and its projects included in the Model Cities area would have lasting influence on the political culture of the city and contribute to the building energy behind neighborhood-level, urban environmental organization. The combination of federal money and local initiative constituted a formative moment in the history of postwar urban environmental activism, unleashing a torrent of creative energy and problem solving that would ripple into the 1970s and eventually throughout Seattle's neighborhoods.[40] The work of MC activists in the Central District articulated in the experience of one neighborhood the environmental, racial, and class inequalities that the postwar metropolitan landscape had helped to produce in the region.[41] In the midst of these interesting times, neighborhood activists offered a holistic and constructive critique of these environmental problems on a local level and solutions to them. Ultimately the Model Cities task force on housing and physical environment—and particularly its focus on environmental health—created clear connections between racial justice and environmental activism. The Model Cities plans, although only partially realized, constituted an early model of sustainability planning on the grassroots level with the help of the federal government.

Toward a Sustainable City

In early spring 1969, the *Seattle Times* published a series of articles that explained the recent organizing energy in the Central District, a neighborhood that was mostly a mystery, or a source of anxiety, to greater Seattle in the late 1960s. For a week each edition of the newspaper devoted lengthy public relations pieces to the priorities that citizens' committees had identified over the previous year as part of the Model Cities plan. The paper's municipal affairs reporter, Don D. Wright, announced that Seattle was to be the "nation's pioneer effort in total urban rehabilitation" and described the city as the "nation's laboratory" in an innovative approach to the "multi-faceted urban crisis which to date has defied solution."[42] Echoing Johnson's speech, he contrasted the program to earlier War on Poverty initiatives, calling it a "concentrated attack upon the ills of the total environment." As the first city to receive federal funding via the program, Seattle found itself at the "the eye of the urban hurricane," according to Wright. In the full-page spread for the inaugural story, the *Seattle Times* featured a large photograph of workers from the environmental health task force traipsing through a "blighted area" in Seattle's Central District, reinforcing the image and familiar rhetoric of natural disaster in the urban environment. Following the typical direction of previous journalistic treatments of urban renewal and urban poverty, Wright introduced Model Cities to Seattle readers as a program devoted to solving the problems of a community in "decay." At the same time, however, his introduction to

Model Cities planning showed the dramatic policy shift from the earlier piecemeal solutions of the War on Poverty. The language that journalists such as Wright used to describe the program—"total environment," "multifaceted," and "total urban rehabilitation"—hinted at features that made Model Cities distinct. The combination of Model Cities' holistic approach and the creative and decentralized planning process it mandated encouraged residents to see various neighborhood problems within a larger social and ecological web.

Two aspects of this comprehensive effort deserve attention. Both are pioneering elements in what would later be termed urban sustainability or environmental justice, and both can be seen in the Fall Drive on Rats, which was emblematic of two major thrusts by which the project directly engaged problems in the urban environment that had roots in poverty and segregation. The physical environment project, on the one hand, and the housing and environmental health projects, on the other, were two of the most significant citizen-driven programs in the Model Cities plan, and their activities and results suggest the contours of an early sustainability movement among citizen and some city planners—one that shared many of the concerns of market preservationists but whose urban priorities differed sharply from those of comprehensive planners downtown. The success of neighborhood design, community-financed homeownership, and better neighborhood planning that MC produced on a neighborhood level in Seattle and elsewhere set the groundwork for more famous and familiar counterculture and neighborhood green activism efforts that emerged in Seattle, San Francisco, and Berkeley. The MC emphasis on self-sufficiency, local economies, better design, and citizen-determined need would eventually provide the watchwords of grassroots urban revitalization during the 1970s.

From its inception, MC meant to shift the emphasis of urban renewal away from demolished neighborhoods, slum clearance, and dislocation. Better physical housing sat at the center of these plans, and the program's creative land use and urban design took new precedence. If racism and inequality were in part environmental issues, housing became the obvious link between the environmental and the social. From the top down, MC meant to concentrate more significantly on preserving and rehabilitating existing housing and promoting homeownership for low-income people while elevating local design standards.[43] Unlike many areas eventually covered by MC projects in other cities, Seattle's MC neighborhood consisted of mostly single-family residences; at the same time, however, 74 percent of the housing there was at least thirty years old when the project began, slightly more than twice the 36 percent for the city as a whole.[44] The neighborhood also had hundreds of vacant or soon to be vacant lots that the project could develop or address as part of environmental health initiatives. Seattle activists, MC staff, and students from the University of Washington Department of Urban Planning began to experiment with a variety of innovative design solutions ranging from new land uses and alternative mortgage financing for low-income families to the rehabilitation of existing structures using Federal Housing Administration (FHA) loans, the creation of green space, and better transportation planning for people in the neighborhood, where automobile ownership was low.[45] On the surface, these aspects of the MC program may seem to be focused primarily on the built environment, yet the program and its proponents saw clear connections between housing and a broader web of environmental and social relationships.

By the late 1960s this web of connections seemed more obvious, undergirded by the work of both ecologists and urban thinkers—including the urban planning faculty at the University of Washington. The groundbreaking books by Rachel Carson and Jane Jacobs, for instance, exemplified such early 1960s works (appearing in 1961 and 1962, respectively), and in 1969, when Model Cities was functioning at its height, Ian McHarg published his highly influential work *Design with Nature*, which became the basis for environmentally based master planning in cities across the country. In Seattle, Richard Haag, the founder of the landscape architecture program at the University of Washington and a practitioner of ecological design principles, served as a local proponent of this work. Just as the UW design students and professors had shaped the market debate, they would lend their influence to the Model Cities neighborhood. As part of a local and national conversation among disciplines, these novel voices provided another context for the more holistic approach to physical planning that people in the neighborhood employed. Professional planners, landscape designers, social workers, and ecologists had begun to think of the environment, therefore, in these more complex and holistic terms by the late 1970s.[46] Students in design programs at the UW, for instance, newly politicized in the struggle for social justice, began to put these ideas into practice. Through the year-long Model Cities planning process, both citizens and urban professionals on loan from the UW, as well as local and county agencies, began to define and articulate interconnections among urban poverty, inequality, and the existing conditions of the natural and built environments.

The *Seattle Times* announced this work of the MC's housing and physical environment task force, carried out in early spring 1969, as the "promise of a decent home in a suitable living environment for every family in the Model Cities neighborhood."[47] Director Hundley took a special interest in housing, explaining that the goal of the task force was to "create innovative housing oriented toward single families primarily."[48] For years the federal pie had subsidized white, middle-class homebuilding in the suburbs around Seattle; MC activists now sought to redirect a slice of that pie, and to secure additional private financing, as a means for building, restoring, and reinforcing the neighborhood's character and scale.[49] Hundley hoped that, in contrast to earlier urban renewal, the new housing programs would "not require displacement of people."[50]

Hundley worked closely with neighborhood residents, including an urban planner named Diana Bower. Like other activists in the neighborhood, Bower had participated in the civil rights movement and was an organizer in the neighborhood's multiracial Leschi area, where she and her family lived at the time.[51] Bower brought design and organizing skills to the process, and during the 1970s she would go on to be a pivotal figure in the International District preservation movement. At the time, she noted the program's flexibility and citizen-centered approach as its strengths. For Bower, too, the social, environmental, and aesthetic aspects of planning were inseparable: "People do not want to live in recognizable 'public housing slabs,'" she explained at the time.[52] Reflecting the MC's emphasis on scale and aesthetics, she noted that even poor people desired "privacy, comfort and beauty on a small scale, preserving the atmosphere of a single family neighborhood." Bower suggested innovative designs to achieve these qualities, including garden apartments, clustered housing units, and the elimination of the five-foot setbacks required by current zoning laws to create more open space for small neighborhood parks.[53]

According to an assessment of the program in the following years, the task force's activities "could be compared favorably to a coiled spring" releasing multiple projects in the neighborhood despite the red tape and slow movement of federal funding for many of the projects.[54] In particular, Hundley and Bower initiated a program that would create cooperatives and condominiums allowing low-income people to access housing and move from "tenancy to homeownership," helping to diminish the low rates of ownership and high number of absentee landlords that MC environmental activists believed contributed to a dangerous physical environment. The crucial element of their plan included a nonprofit housing development corporation called the Housing Development League. Two years after the program was announced, over six hundred housing units were under consideration for the neighborhood, and along with a land bank, the program had become the dominant force in shaping housing for the MC neighborhood's residents.[55]

In addition to contributing a new design and local financing for single-family homes, the program concentrated on rehabilitation of existing housing. Rehabilitation could mean demolishing structures in the neighborhood, many of which were abandoned, dilapidated, or condemned, to make way for the new. Implementing the program entailed destroying 1,030 structures after 1968, but it created almost the same number of new units by 1971.[56] The program's visible effects in the neighborhood revealed a commitment to rehabilitation that echoed the desires of activists, designers, and even city officials to preserve older buildings for reuse or to maintain what Bower called "atmosphere." The target of these rehabilitation programs included hotels and significantly aging buildings in the International District and in Pioneer Square, which had recently been designated a historic neighborhood. In this way, the same energies expended on historic preservation at the Pike Place Market seemed to operate in the less-beloved neighborhood with far less fanfare. In coordination with CAMP, the rehabilitation program hired and trained young people from the neighborhood and created a "fix it wagon" operated by trained workers, also from the neighborhood, to perform repairs on homes purchased through the FHA program.[57]

While Hundley and Bower's housing and environment task force took on this more macro-level focus in the holistic planning process, the environmental health program attacked the more obviously ecological aspects of the neighborhood's environmental needs. But activists in this realm found these needs at the intersections of human bodies, physical spaces such as the home, and a sometimes dangerous nature.

Community Sanitation Aides' field training, Oct. 1, 1970. Model Cities Program Environmental Health Project Records, Photograph Collection item no. 69300. Courtesy Seattle Municipal Archives.

Isaac Banks was a pioneer in this effort. A longtime resident of the Central District, Banks directed health services for the Seattle–King County Department of Health when Walter Hundley hired him to coordinate the "Environmental Health–Rat Control Project" for Model Cities. The project changed its name to the "Environmental Health Project" by 1969 and soon thereafter began its baseline study of the neighborhood. In an interview with the *Seattle Times* real estate editor on taking his position, Banks described his work in the program. Reflecting Model Cities' effort to empower neighborhood residents, he said, "This is the people's program," explaining, "They must assess their problems, establish a program, and bring their needs into focus." Those needs were defined holistically from the start as Banks began to implement the project in the neighborhood, where he brought county health department skills to bear on the more concentrated effort in the Central District. "All the problems are interrelated,"

Signing of the cooperative cleanup contract between Model Cities Central Area Motivation Program and the Environmental Health Project (Walter Hundley, center),
Mar. 31, 1971. Model Cities Program Environmental Health Project Records, Photograph Collection, item no. 69294. Courtesy Seattle Municipal Archives.

Banks explained. "You can't build up health unless you can improve living conditions, and housing." He added, "Environmental improvements
are dependent upon improvement in the employment field."[58] At each step
in the planning process, organizers and activists such as Banks kept these
principles and a picture of environmentalism, social justice, and human
bodies in mind as they worked their way down the alleys of the Central
District.

Coming to MC from a public health background, Banks adopted an approach reflecting this broader conception of environmentalism, which
emphasized the science of ecology and the dynamic between human beings and the physical environment. Indeed, Banks and many others public health workers saw themselves as the vanguard of the environmental
movement in 1968. The King County Health Department's annual report for
1969 reinforced this conception of the environment and expressed the point
of view clearly within the context of the general public's rising interest in

ecology. Banks's home agency even claimed credit for keeping the flame of environmental activism alive before it was popular. "As the 1960s fade into the smog we look to clearer horizons in the 70s," the department's director wrote on the eve of Earth Day in 1970. "Suddenly it seems protest against pollution of environment has become fashionable. The public has discovered the word ecology." Yet, he remarked, "Until recently environmentalists in public health have likened themselves to voices crying in the wilderness. Alarmed, the public is now joining ranks to prevent our environment from becoming a wilderness!"[59] The picture of the urban environment as a wilderness to be tamed informed the worldview of the health department and ultimately influenced the perspective of MC workers who would comb the alleys for rats and refuse. "Public health people," the report explained proudly, "have taken positive steps to rehabilitate our environment," working to "protect our food and drink, our places of work, play and rest."[60]

Beginning with this orientation, the Environmental Health (EH) Project helped to define—or redefine—the meaning of the words *environment* and *ecology* as specifically urban, placing concern for people and poverty squarely at the center of a larger environmental discourse coalescing in Seattle. The program did this by developing public awareness through outreach, education, and job training. The vigorous role Banks and other public health workers played was always on display in the press and in public forums, for these organizers and activists aggressively advocated an environmentalism that could be applied to urban situations and specifically the MC neighborhood.

Banks and his staff effectively used publicity. According to one reporter at the time, when Banks discussed conditions in the MC program with the press, he "pull[ed] no punches." Banks explained, "I've seen where rats have urinated on eating utensils; where cockroaches have crawled from commodes to milk bottles."[61] Such quotable material was effective, creating a picture of natural dangers, but so were the public appearances that the Model Cities program workers made during the program. But these images and activities could seem to stigmatize the neighborhood, too, reinforcing a kind of environmental determinism to poverty for a general public. Still, Banks's urban environmentalism challenged the status quo, making a dangerous environment a source of empowerment and critique. At one "Earth Week" event at Westlake Center—a crossroads in Seattle's downtown business district and blocks from the Pike Place Market—MC workers erected a large photo display about their vermin control and eradi-cation program alongside displays by the likes of the Sierra Club and other mainstream environmental organizations. During an Earth Day event downtown in April 1971, the Environmental Health Project displayed their rubbish-collecting trucks emblazoned with the Model Cities logo. Accord-ing to their activity report, a "Project Training and Education officer" stood next to the display to explain "vector control" efforts to a curious public. In addition, after 1970, the program made a general effort to reach out to the community by radio and television and at public events and schools. Yet it was the organizing work in the neighborhood that would make all the difference for a community in which many residents remained suspi-cious of government and social planners.

As Banks's staff and committee assembled their early grant proposals for Model Cities funding, they met often with citizens in the Central Dis-trict. Throughout 1968 in weekly planning sessions they worked with resi-dents to define "widespread environmental insults to health and well be-ing." Community members were not always welcoming to the professionals and activists during these often turbulent meetings. In a remark that sug-gests the level of distrust and the past abuses of federal urban renewal schemes, the MC staff noted that some people in the neighborhood were "hostile toward city officials" who, they claimed, were "not responsive to the needs and complaints of the Model Neighborhood area." Echoing the arguments of the *Afro American Journal*, some neighborhood residents felt that MC was a "hoax" and that the "planning meetings were simply talk sessions to pacify Model Cities residents during the 'hot summer' and that no meaningful improvements [would] come to the 'ghetto.'"[62] Yet of all the Model Cities task forces, the one on environmental health would show the most tangible results within the neighborhood and elsewhere in the city. Following the Model Cities goals to create "a citizen-planned program," from 1967 to 1974 environmental health workers made yearly rounds of the MC neighborhood, interviewing residents, filling out surveys of problems in the physical environment, and then returning to reassess the program's success. Together the EH staff and the neighborhood defined, for instance, everything from the "natural deficiencies" of "topography and drainage" to the conditions of parks and playgrounds and the problem of pests. At every point in the process, the workers attempted to maintain "high vis-ibility and impact" in the area.[63] The series of EH reports helps show how residents viewed their environment in the late 1960s, how the MC workers partly shaped that view, and how together they set priorities for solving the environmental issues facing people in the community.

For good reason, danger and safety seemed to preoccupy this citizen-driven vision of the neighborhood and environmentalism. Simply dropping the mention of "rat control" from its title, for example, shows the program moving toward a broader vision and emphasis on a variety of environmental hazards.[64] The inhospitable environment of the MC neighborhood included noise and sound pollution, noxious weeds, abandoned cars, vacant lots, and burned-out houses, as well as infestations of cockroaches, silverfish, mosquitoes, and flies.[65] Yet rat infestation and harborage, the primary focus of the health department workers in the city for some time, continued to top the MC program's list of concerns, for it constituted an important symbol of the environmental problems the community faced and the holistic remedies they required. The rat became a kind of indicator species. To control rats in the neighborhood, according to a 1970 report, would have required "improved garbage storage" and a solution to the problems of "accumulated litter and rubbish, dilapidated garages and sheds, abandoned vehicles, unimproved alleys, unkempt vacant lots, overcrowded and substandard housing units, etc."[66] Perceptions of endangered human health and landscapes helped to mobilize people in the neighborhood to take control of space individually and collectively.

The Environmental Health Project made critical links between neighborhood planning and environmental dangers and deficiencies. Rats and rat habitats provided a lens through which to view the broader problems of the neighborhood's overall environment. Absentee landlords and the high turnover rates of low-income renters in the area, for instance, as well as the sheer density of people and their waste, contributed to the situation. "Because of the low care and maintenance given to the property by the landlord," the report noted, litter was "left to accumulate."[67] Rats plainly revealed the inseparability of environmental problems and poverty. Through outreach, Banks and his crew helped people inside and outside the neighborhood to see the relationships among their bodies, the environment, and the legacy of segregation that had contributed so clearly to this dangerous environment.

During 1970 and 1971, as Banks and his EH aides began to define a vision of interrelated environmental and social problems, the seemingly

Environmental Health Project Community Sanitation Aide Supervisor spraying a Model Cities neighborhood residence for cockroaches as an example of vector control activities, Mar. 31, 1971. Model Cities Program Environmental Health Project Records, Photograph Collection, item no. 69287. Courtesy Seattle Municipal Archives.

endless neighborhood meetings began to bear fruit. Having completed the baseline study of environmental problems in the neighborhood, program members began to attack them. Each quarterly report of the EH program featured photographs that documented its successes. Before-and-after images showed young men from the neighborhood dismantling collapsed garages, clearing blackberry brambles from overgrown vacant lots, and loading Model Cities trucks with garbage and abandoned appliances. Like a low-budget, 1970s version of the WPA photographs documenting workers in the Great Depression, the pictures advertised the local progress of putting young people to work. The report for 1971, when the newly elected president Richard Nixon suggested canceling the program, showed its frantic momentum. That year EH staff labored alongside CAMP workers to clear 499.33 tons of garbage, rolling through the MC neighborhood with four new trucks throughout the spring of each year. Workers documented the contributions of neighborhood groups involved in these mo-

Environmental Health Project personnel loading a locally owned truck during neighborhood cleanup, Oct. 1, 1970. Model Cities Program Environmental Health Project Records, Photograph Collection, item no. 69308. Courtesy Municipal Archives.

bilizations. During 1970 the list of participants ranged broadly, including local high school students, CAMP workers, and members of the Chinese Community Chamber of Commerce, among many others. And during 1971 the EH project reported that "citizen participation" had been "direct and effective." After EH workers returned that year to evaluate the strides they had made in the neighborhood, they remarked, "The degree of improvement observed through this restudy was indeed gratifying." The project reported the assistance of "various local community councils also involved in improving local neighborhood conditions."[68]

While EH workers identified and expanded the range of environmental dangers to be addressed, they also expanded efforts to educate the public about controlling sources of disease. It would be difficult to assess the impact of these efforts to shape public opinion, but certainly MC laid a foundation for people in the community—especially in public schools—to consider themselves part of an ecosystem, even if it was represented as po-

Men and one of the cleanup trucks in action in the Model Cities neighborhood, June 30, 1971. Model Cities Program Environmental Health Project Records, Photograph Collection, item no. 69278. Courtesy Seattle Municipal Archives.

tentially dangerous. Each quarterly report detailed these activities and attested to the broad range of subjects that fell under the rubric of environment and health. Environment could be inside or outside the home; it could range from "community planning" meetings with adults to "home safety" meetings with senior citizens. Typical meetings might reach a crowd of thirty to fifty people. During February 1970, for instance, EH workers spoke to fifty-seven adults at the Japanese-American League about environmental health. At the Cherry Hill Community House meeting in June, the topic was "ecology." EH workers also showed up at public events, such as the Pike Place Market Fair in the summer of 1970, and in November they spoke to a Boy Scout jamboree about "insect and rodent control."[69] During 1970 the program conducted a series of "neighborhood block meetings" and reported contacting over "650 residents" with "mini teach-ins in schools, churches and community centers," as well as educating "residents in the community on a person-to-person basis."[70]

CAMP worker interviews Model Cities neighborhood resident on her knowledge and acceptance of the cleanup project, July 1, 1970–Sept. 30, 1970. Model Cities Program Environmental Health Project Records, Photograph Collection, item no. 69317. Courtesy Seattle Municipal Archives.

In compliance with the MC's federal guidelines, which stressed a holistic approach, Banks and his staff attacked problems of the physical environment as both social and physical ills. Training and jobs were critical to the overall process of solving the community's environmental problems. If poverty and segregated labor markets—resulting in degraded housing and poor conditions—were causes of the dangerous environment in which so many MC residents found themselves, the EH program would help to address this cause as well. One activist during this period argued, "As Model Cities grew we made sure that Model Cities and everything we did was integrated. When funds were used they were being used to integrate and desegregate wherever the tentacles of Model Cities went out."[71] Environmentalism, as defined by MC activists, therefore did not separate ecology from the social goals of the larger civil rights movement.

Consequently, one of the project's objectives for 1970 was "to recruit, employ, and train Model Neighborhood residents as environmental health workers."[72] The designations given to these workers indicated the health department's continuing influence. Reports from the period thus referred to the employment and training of "health aides" or "sanitation aides" in part because the Model Cities organizers considered the recruits' training in ecology to be paths to employment in the health department or other related municipal and county agencies. Again, this training and employment was crucial to the project's vision of engaging the whole environment, social and physical. MC workers described the "Model City resident employees" as "change agents."[73] As with every aspect of Model Cities, employment and desegregation were critical to the larger process. To that end, the EH program emphasized the importance of "having a *career ladder* for these Aides, training . . . these Aides so they would be *prepared* to pass *civil service examinations* if that should be required at the time that funding to the Model Cities program diminishes."[74]

Between 1968 and 1971, the EH program employed hundreds of laborers in the neighborhood to perform surveys, remove tons of refuse, and perform "surveillance of overgrown vacant lots," part of the program's commitment to systematic training of "indigenous Community Sanitation Aides." Throughout the late spring and early summer of 1970, community sanitation aides logged hundreds of training hours learning about such topics as the biology and habitats of the Norway rat, the significance of flies to public health, and insecticidal equipment. They also attended classes in leases, mortgages, and landlords. This training, with its overlapping attention to the natural and built landscapes, helped to mobilize citizens to take control of their environment from multiple points of entry. At the same time, it helped identify and push forward viable candidates for more formal training in local colleges.[75] In April 1970 the program announced that four of the aides had been accepted in colleges.[76] By 1971 the EH program had expanded from its earlier emphasis on rodents and a dangerous environment to study the potential for lead poisoning in MC housing. According to the program's preliminary reports, as much as 90 percent of the housing posed a threat. With MC funding, the program sent environmental aides into schools to test children for lead poisoning; 10 of the first 200 children examined tested positive. The program also began coordinating with the humane society to round up stray dogs in the neighborhood.

As part of its training activities, and in an effort systematize the work, the EH helped to create a nonprofit neighborhood group to oversee the popular and successful cleanup program. The group, calling itself "Pack Rats," constituted one of the city's earliest citizen-driven nonprofit programs that explicitly linked neighborhood cleanup and recycling efforts to the larger goals of the mainstream environmental movement. The program also produced a graded list of "environmental specialties" that essentially reproduced the "environmental specialist ladder" the EH project had created for its trainees.[77] The grades, which were designed to aid in preparing workers for civil service, also displayed the range of activities and work training that EH and Pack Rats helped to sustain in the neighborhood. "Environmental specialists," for instance, worked crushing cans and operating cardboard baling machines in neighborhood recycling centers, while an "area supervisor, environmental operations," grade 8 might educate "the general public in environmental and ecological areas through use of media."

Operating out of the "Pack Rats Nest," on Empire Way (today's Martin Luther King Way) in the Central District, the company adopted a letterhead and logo featuring a rat holding the Earth Day ecology flag and asking, "Have you thanked a green tree today?"[78] The business, which appeared to be struggling to survive after the Model Cities period, nonetheless exemplified an early attempt by Seattle activists and small business owners to capitalize on the new neighborhood environmental consciousness and encapsulated the local and integrated ethos nurtured by MC. The group sought a wide appeal: "Pack Rats are People! Black and White, Old and Young, Students, Drop-Outs, Long Hairs—All With A Purpose—To utilize all recyclables available, to control solid waste environmental pollution, and to create jobs and opportunities for as many people as possible." The influence of MC principles could be seen in the bylaws of the Pack Rats, whose mission was to improve the "environmental and ecological quality of the Greater Seattle and Western Washington communities through solid waste recycling," to provide "ecological education of young people and the general public," and to create "employment opportunities for young people and disadvantaged low income persons." Pack Rats created a context for employment in the community and showed how the influence of MC rippled into the private sector and public consciousness as environmentalism became a popular movement. As the Model Cities program later expanded to the entire city in the 1970s, these ideas would gain more traction among other neighborhood groups and activists who learned from its experiments.

The "Fall Drive on Rats" marked one part of a much broader and fluid effort to address urban environmental concerns in the Model Cities program.
The program in Seattle offers an overlooked example of this shift toward a
more inclusive definition of environmentalism and its origins in the confluence of federal programs and the local initiatives of civil rights activists.
With its mandate for comprehensive approaches to urban environmental
problems, the Model Cities program mobilized a variety of citizens around
concerns for their environment and developed and reinforced an existing
sense of place and collective power within the MC neighborhood. The Central District activists made explicit links among environmental problems,
poverty, and racial justice. By the late 1960s and early 1970s, Model Cities—
the program's successes, failures, and unintended consequences—helped
to inspire a wide-ranging citizen movement in the city at a moment of economic flux. By the 1970s, during the Nixon administration, the program
had expanded beyond the Central District to encompass projects throughout the city. It gave shape and inspiration to the neighborhood-based politics of that era while spurring a variety of urban environmental projects,
organizations, and actors, including a growing counterculture green movement. Indeed, the eventually successful activists fighting for the preservation of the Pike Place Market, who defined the place as a living, breathing neighborhood, echoed the similar efforts of Model Cities activists. The
market fight made sense only against the backdrop of the War on Poverty
and the Model Cities program, which sought to empower poor people and
address failures of postwar urban renewal and civil rights protections on
a neighborhood level. Victor Steinbrueck and his followers drew from this
larger antipoverty and civil rights climate in the city during the late 1960s
for explanatory power. Model Cities, with Seattle as its first funded project
in 1967, was a critical component of this emerging urban sensibility. Model
Cities may have been the last gasp of the War on Poverty, but it continued
to influence place making in the city well into the 1970s and 1980s.

A measure of the program's success in mobilizing creative and dispersed experiments in community renewal and in nurturing a new urban
environmentalism during the late 1960s can be found in its extension during Richard Nixon's first administration. Nixon, who took office in 1969,
was no friend of the civil rights movement or the War on Poverty. He and

many conservatives in the new government associated Model Cities with the failures of the Great Society. According to one historian of this period in U.S. urban policy, however, the program at least theoretically appealed to Republican ideas of small and local government.[79] Noting its success in Seattle and elsewhere, the Nixon administration decided to expand the program in several test cities as part of the "planned variations" program. Planned variations would test the new administration's ideas about the efficacy of block grants, the urban funding program that would eventually replace Model Cities in 1974. As the test run for Nixon's new urban policy, planned variations would strengthen mayoral control, reduce federal oversight, and include the entire city in its purview.[80] The effect of these decisions at the federal level would be to temporarily raise the level of available funding for the program, increasing competition among people, creating more projects, and (inadvertently) spreading the spirit of grassroots urban renewal throughout Seattle. But the Central District finally lost out.

The local press noted these changes to the program in 1971, when Mayor Wes Uhlman announced the new federal program to the city council. Seattle was chosen as one of sixteen cities to receive a new $5.2 million grant, the funds to be spent on expanding its Model Cities program to three new neighborhoods. Uhlman also announced that Walter Hundley would be the director of the new citywide program.[81] In September Hundley told the *Seattle Post-Intelligencer*, "Under this plan all areas of the city described officially as blighted would be eligible for participation in Model Cities," including physical and social rehabilitation.[82] The expanded program would comprise areas south of the original MC boundaries: the Georgetown and Highpoint neighborhoods and "surrounding blighted neighborhoods," as well as parts of Ballard and the Fremont neighborhood near the shores north of Lake Union.[83] These fateful decisions about spreading the wealth and organizing energy of MC would contribute to the building neighborhood renewal energies in the mainly white, middle-class north end of the city. At the time some critics noted that the program would give the mayor too much power, a problem that the original Johnson-era program hoped to address. Yet according to the Central District's newspaper, the *Medium*, the community overwhelmingly supported the program's expansion.[84] By October 1971, the new citywide program was up and running. The *Seattle Times* paraphrased Hundley: "It appears the program is being accepted as well or better in the predominantly-white new areas as it was in the mostly-minority race original area."[85] Certainly that year would be a turning point for Seattle's neighborhood activism as more Seattleites sampled a taste of community organizing. As the 1970s wore on and the post-Vietnam and local recessions sunk in, the federal funding would ultimately dry up, but the spread of Model Cities would have lasting effects on the physical and social landscape of Seattle, influencing neighborhood activists to think in new ways about remaking the city from the bottom up.

Emerging against a backdrop of failure in urban renewal by the late 1960s, Model Cities created the context for an expansion of a variety of grassroots projects, many of which reinforced a strong sense of community and a sense of power for citizens who sought to control the production of space in the city. The grassroots movement that challenged developers' plans for the Pike Place Market had framed their argument to accentuate the ways in which urban renewal and federal antipoverty spending had been hijacked by downtown developers. Such projects, they argued, succeeded at the expense of historically important, living neighborhoods and the poor people who lived in them. The activists in MC made similar arguments as they turned ideas of blight and danger in their environment to their favor as part of a comprehensive critique of postwar urban policy priorities and segregation. Together, these critical episodes in Seattle's early attempts to negotiate the meanings of the sustainable city helped to bring social justice, preservation, health, the built environment, and scenic amenities to the center of the discussion of urban environmental problems. Human bodies and human habitats became part of this public discussion of urban environmentalism.

In this process, men such as Hundley would act as transitional figures, brokering important initiatives with the city while building a bridge between the earlier civil rights era and the more urgent voices of the movement that followed. Ultimately, both Model Cities and the Pike Place Market dispute taught all Seattleites new ways of viewing their urban environment while encouraging grassroots activists to link the production of space to their politics in concrete and immediate ways. Despite the achievements of Model Cities in the neighborhood, the late 1960s and early 1970s would remain a tense period for race relations within the city. The period would also see the emergence of a bolder and more confrontational set of actors who would take the lessons learned in Model Cities beyond the confines of one neighborhood.

3

Open Space

Before daybreak on a March morning in 1970, a large group of men, women, and children made their way along a Puget Sound beach, just north of downtown Seattle. Braced against the chilly salt air, the figures moved past darkened windows of beach houses in the ritzy Magnolia neighborhood. Their destination was a U.S. Army base, Fort Lawton. A 1,100-acre military outpost established in the late nineteenth century, the compound sat atop a promontory with a view of the Olympic Mountains to the west, only eight miles from Seattle's high-rises and new freeways. Beach homes gave way to a steep hillside as the group crossed the boundary dividing residential from military territory. They stayed close to a slippery clay wall below the bluff, the high tide lapping at their feet. Then they found their trail and began the difficult climb. Reaching a ridge below the rim of the bluff at around 5:30 a.m., they waited.[1]

This scene was no typical predawn Seattle nature walk but an invasion, a charged symbolic act of an organization calling itself the United Indians of All Tribes (UIAT).[2] The morning walkers purposely trespassed on government land to make a political point and to claim a piece of urban real estate. A reporter from the *Helix*, Seattle's leading underground newspaper, accompanied the protesters that morning and described the next moments: "They wait huddled together, munching on chocolate to await the moment of attack which has been set for 6:30. Just before this time we hear a motorized MP patrol pass by and when this has faded into the distance the Indians break their cover and move quickly into position. A small teepee is set up and a warming fire is built. Full of high spirits the Indians dance and chant to celebrate their victory."[3] The Indians then unveiled a proclamation: "We the Native Americans re-claim the land known as Fort Lawton in the name of all American Indians by the right of discovery." The old fort, they argued, was "more suitable to pursue an Indian way of life": "By this we mean 'this place does not resemble most Indian reservations.' It has potential for modern facilities, adequate sanitation facilities, health care facilities, fresh running water, educational facilities, fisheries research facilities and transportation."[4] In making their claim for an "Indian way of life" in the city, the UIAT also contrasted their vision for a livable Indian city to conditions prevailing in contemporary Seattle, a city with "no place for Indians to assemble and carry on Tribal ways and beliefs." Along with announcing plans to develop a center for Native American studies and a "great Indian university," the group proposed an Indian center of ecology to "train and support" Indian youth in "scientific research and practice" aimed at restoring the "lands and waters to their pure and natural state."[5]

The UIAT blended media savvy and effective irony with direct confrontation, a style increasingly common after 1968 among antiwar activists and one apparent during the famous Alcatraz occupation a year earlier. The group purposely evoked frontier clichés. Powerful images of Indians taking federal land could not be denied. Jane Fonda even arrived to show her solidarity and be photographed with the activists. The dramatic occupation made an irresistible story for Seattle's press, and the events—and continuing protest and court battle—drew wide attention to Seattle's youth and environmental movements. The possibility of reclaiming—or "liberating"—such a large chunk of urban real estate from the federal government at the height of the Vietnam War era resonated with many Seattleites in the time's heated political atmosphere.[6] Like the Alcatraz protest in 1969, the Fort Lawton events highlighted injustices that Indians had experienced (and continued to experience) nationwide in rural settings and especially in West Coast cities, including San Francisco, Oakland, and Seattle.

Figures in the story of environmental activism in Seattle did not always stick to expected scripts. On the bluffs above Puget Sound the UIAT juxtaposed clichés with concrete political goals, both drawing from and confounding dominant stereotypes about Indian life: Indians could be both urban and ecological; they could be both warriors and savvy urban political operatives; and they could evoke timeless spiritual connections to the land and demand crucial "modern facilities." Most important, drawing from the neighborhood organizing underway on the poorer south side of the city, they explicitly folded conditions of urban inequality into their environmental rhetoric. At the fort that early spring the Indian activists claimed a physical piece of green space, but just as important, their high-profile protests had claimed political space in urban politics in the late 1960s. The rhetoric and style of their protest may have echoed the Red Power movement. Their activism may have evoked the horrible conditions of rural reservations and the reality of violated treaty rights in Washington State. Yet their environmental rhetoric did not merely manipulate stereotypic images of Indians as stewards of the land. Rather, spokespersons such as Bernie Whitebear (Colville) and other activists at the fort drew inspiration and tools from the specific, local, multiethnic civil rights movement in the city, especially from the lessons learned from environmentalist community organizing in the Model Cities program. Whitebear and his comrades—including Black Panther Party and Asian American activists from the Model Cities neighborhood—represented a younger generation. The UIAT used environmental rhetoric in their protest, playing up Indian stereotypes, but their deeper, political goals echoed the planning, sanitation, and health concerns of other urban environmental activists hard at work in the Model Cities neighborhoods, including the Central District, the International District, and Pioneer Square. Their demand for modern facilities, sanitation, health care, and transportation may have appeared dissonant against their use of stereotypical imagery, but these environmental issues were the shared concerns of many poor people and people of color in a city just emerging from segregation. The UIAT's plans for the fort effectively announced their presence as a stakeholder in the city's shifting political, environmental, and spatial order.

The invasion, however, interrupted another vision for the land. Although expressed suddenly and dramatically at Fort Lawton, the desire to liberate urban land (especially the fort) for new uses had percolated among various mainly white, middle-class civic groups in Seattle for some time. Uniting affluent neighbors, outdoor groups such as the Seattle Audubon Society (SAS) and Sierra Club, mainstream civic organizations such as the League of Women Voters, and prominent pillars of the community in a loose coalition, the Citizens for Fort Lawton Park (CFLP) had pressured the federal government to donate the fort to the city beginning in the early 1960s. Many of these groups and individuals were part of the preservation movement opposed to the Pike Place Market demolition, but the CFLP also included some who supported the downtown developers hoping to remake the city. In the late 1960s they found common cause in preserving nature in the city. By the time the Indians claimed the fort "by right of discovery" in 1970, the citizens group—and especially women active in the SAS—had all but obtained the "open space." The Indians, therefore, directly challenged the coalition's plans and negotiations at a sensitive moment.

But most important, the standoff represented another clash between two sets of political sensibilities and urban priorities at the end of the 1960s, suggesting the outlines of a shifting political order that once again put the environment and the production and control of urban space at its center. The "battle of Fort Lawton" highlighted obvious fissures forming within postwar liberalism during the Vietnam War and between mainstream liberals and an increasingly boisterous youth movement. The meaning of nature, and even what constituted the environmental movement, sat at the center of these conflicts. The events therefore provide a critical urban spatial and political frame for understanding the nascent post-1960s environmental movement in the United States. Mainstream groups, such as the Audubon Society, supported a more conservative position, seeking to preserve and protect traditional ideas of nature and even "wilderness" in the city. The UIAT offered an alternative social, environmental, and political vision for the space that echoed the holistic approach learned in the Model Cities program. Both groups spoke the language of nature, beauty, and even ecology—and they meant it—but they were also describing their stake in a social and political order, sometimes at each other's expense.

The UIAT sought an exclusively Indian space within the city, one that would offer social services, a space for cultural identity, and access to nature. The Citizens for Fort Lawton Park and SAS envisioned a similarly idealized notion of "open space," but they argued that Fort Lawton—and other spaces like it—could alleviate social and environmental ills and bring order to a chaotic urban environment. Nature, they believed, might provide a safety valve for the energies of the urban youth—those in the UIAT—who now appeared to threaten their plans for park. And both groups faced the U.S. Department of Defense's plans to build an anti-ballistic missile (ABM) facility at the fort. Urban Indian activists, the CFLP, the SAS, and the federal government, then, participated in a complex public negotiation about the potentially powerful meanings of nature, order, and social change in the post-1960s city.[7]

This debate hinged on identifying who defined the terms of public discourse about nature and urban problems. Just as the Pike Place Market debate and Model Cities program exhibited both similarities and sharp contrasts in urban priorities, the different meanings that Indians or coalitions such as the CFLP brought to their postwar environmental activism grew not just from the heady times but also from the material conditions and politics of metropolitanization.

Freeways and urban renewal, suburbanization and downtown's decline, as well as class, ethnic, and gender conflicts emerging during the civil rights era—all these gave new form and urgency to a variety of environmental critiques of the urban landscape.[8] How and why, in other words, did different urban actors—including Native Americans, bird-watching women, white hikers, and neighborhood activists—become politicized around nature and the urban environment, forming an active base for a new movement, even as the meaning, vocabulary, and aims of that movement remained amorphous and contested? The Fort Lawton dispute was not just a disagreement over public space, then, but a public discussion concerning what and who counts as the public. Market preservation advocates had employed ideas of race and class in their arguments to preserve the market, though in a muted and mostly indirect form. The Fort Lawton clash more directly revealed the postwar urban environmental discourse to have become a product of the modern urban landscape and its multiple political tensions by the late 1960s. The events around Fort Lawton—a fight for green space and a fight for political space—tell the story of this local environmental consciousness within the larger national drama of the late 1960s.[9]

Although environmentalism appeared to have achieved wide acceptance by the first Earth Day, in 1970, the diverse activists who derived power and purpose from defining open space in the city or from confronting urban renewal or federal impositions, or in organizing their own neighborhoods, often shared a loose vocabulary that included words such as *blight, beauty, wilderness, ecology, preservation,* and *nature.* But they used this language to explain and justify very different and sometimes mutually contradictory political goals. Such words had power, and each group used them to seek a foothold in a dynamic postwar cityscape. The most ubiquitous phrase of the era—*open space*—was up for grabs and could be filled with multiple meanings. By the end of the 1960s, an emerging urban environmentalism occupied the unstable ground of urban politics, a context that ultimately could give shape, power, and meaning to this movement.

Fort Lawton's Origins and the Search for Order

Fort Lawton was born in the Progressive Era with the explicit mission of bringing order to a young and disordered city racked by violence. In the mid-1880s the local Chamber of Commerce persuaded the federal government to build the fort on Magnolia Bluff in part as a defense for Puget Sound's military assets and to bolster Seattle's prestige, but the fort had a more important domestic purpose. Local business leaders' desires for a secured metropolis coincided with a nationwide economic depression in the 1880s, labor unrest, and a surge of anti-Chinese violence in Seattle and Tacoma. Many believed an unruly city threatened the region's evolving social and economic order.[10] "On account of its contiguity to the coal mines, its central location and ease of access," the county commissioner reported in 1886, the city was "likely to be chosen as the theatre for lawless demonstrations and outbreaks by all the disorderly elements within reach."[11] When the army surveyed potential sites around Puget Sound in 1894, the secretary of war agreed, writing, "The vicinity of Seattle offers the most favorable conditions for location of a post" because "the future use of troops will be demanded" for a city where "exhibitions of lawlessness beyond the power of the state to control have so frequently manifested themselves."[12] Seattle's Chamber of Commerce was so eager for a fort in the city that it rallied support to buy 700 heavily timbered acres on the bluff and gave the land to the federal government. The secretary of war approved the land transfer in March 1896, and the city handed over twenty-seven separate parcels amounting to 704.21 acres.[13]

From the start the fort was both a symbol of and, given the military force housed there, a power for order. Once the government had gained title, however, it was slow to develop the land as the business community had hoped. Perhaps disheartened, city leaders turned their attention to a less overt form of social control in the young city: parks. In the Progres-

Aerial photograph of Fort Lawton from northwest, ca. 1940. *Seattle Post-Intelligencer* Collection, image no. 2002.48.823. Courtesy Museum of History and Industry.

sive Era parks played a powerful role in urban design, a force for shaping ideal subjects and sensibilities. From 1897 to 1907 the well-known Olmsted brothers, John and Frederick Jr., sons of the famous landscape architect Frederick Law Olmsted, who created New York's Central Park, designed a system of boulevards and parks for Seattle. Early on, then, the green spaces that became part of the Olmsteds' comprehensive parks and boulevard plan—including Green Lake and Seward Park—began defining the way many Seattleites experienced nature in the city.[14] As in other cities, such parks would help demarcate the city's upper- and middle-class neighborhoods.[15] During this period, Seattle's parks commission hired the Olmsted firm to redesign Fort Lawton for inclusion in the extensive park system.[16] The proposed plan marked the fort as an outer boundary of the city's park plan, cutting off what the report called "wild country" from an emerging pastoral suburban ring around the inner city.[17] From its inception, then, civic leaders had hoped that Fort Lawton might give shape to Seattle and that, as the prevailing ideology at the time maintained, nature would help bring a sense of calm and gentle order to the young frontier city. Some of these ideas would endure in the conservation movements in the second half of the twentieth century.

Yet Fort Lawton's inclusion in an overall city park design never materialized. After the Progressive Era and the high-water mark of park building in Seattle, the nationwide clamor for scenic parks subsided. The most notable early use of the fort came during the Spanish-American War, when the undeveloped surroundings were used to stable horses for regiments on their way to the Philippines. By World War II and the Korean War, though, Fort Lawton had become genuinely active, reflecting Seattle's overall mobilization for war. Later the fort hosted a radar station and, in the 1950s, a control center for Hercules and Apex missiles.[18] The army had a presence in the city, but the fort never fulfilled the original hopes for an installation on par with the Presidio in San Francisco, as boosters had hoped. Nor would the Olmsteds' more elaborate plans for the fort be realized. The fort's acreage mostly served as an informal preserve for the surrounding neighborhood, a place to walk dogs and enjoy scenic beauty. As suburban development exploded on the city's periphery after 1950 and as industry and people moved beyond older city limits, a variety of urban activists during another period of urban flux and unrest, especially the conservationists of the 1960s, once again eyed the massive undeveloped land near the heart of the city.

Contexts for Middle-Class Protest

The shifting physical landscape of metropolitanization, as well as its shifting urban priorities, formed the context for a variety of activists who eventually mobilized around Fort Lawton.[19] The contested meaning of the proposed open space there reentered the realm of public debate at the same time as the market controversy blossomed and as Model Cities activists grappled with the quickening pace of postwar changes. Like their counterparts in Allied Arts—especially leaders, such as Victor Steinbrueck—an emerging group of Great Society–inspired beautification advocates began to challenge the most obvious examples of these regional changes: the new freeway system and locally organized urban renewal downtown. Although these early local voices of dissent did not bear directly on the Fort Lawton controversy, they, along with a growing critique among American professionals and academicians, offered an important precedent, set of tools, and rationale that open space advocates in Seattle employed.

As local activists began to agitate from below, the federal government took a leadership role. In this conversation about aesthetics, morality, and environment, the idea of "open space" became an almost metaphysical solvent for urban problems. The phrase was ubiquitous in the 1960s, and like *blight* and *beauty*, it was hard to define. Yet experts, policy makers, and President Johnson used the words often to describe a new way of thinking about the American landscape in the postscarcity age. In his 1965 "White House Message on Natural Beauty," President Johnson called on activists such as those in Seattle to "restore what [had] been destroyed and salvage the beauty and charm of our cities."[20]

Ann L. Strong's *Open Space for Urban America*, a publication produced for Johnson's Department of Housing and Urban Development in 1965, exemplified the coalescing conservation and open space movement of the 1960s. Her work directly influenced the park advocates at Fort Lawton. Written for the "informed layman and State and local officials," the widely distributed publication described new conservation's goal to create an urban society that would retain "a continuing association with the natural world." Reflecting contemporary assumptions about the affluent society of the 1960s, Strong advocated "an orderly, efficient, and attractive urban environment within which men may enjoy the good life."[21] These government publications codified an ideology of public space, but they also gave examples of successful projects and advertised the federal role in open space preservation. *Beauty, blight,* and *open space* formed a persuasive vocabulary that the SAS and the UIAT could use to guide the taste and aesthetics of local communities or to advance other agendas. Open space offered a brake on development, a relief from the relentless changes that faced the United States after World War II. This amorphous concept and alluring idea would become a crucial part of the language of the modern urban environmental movement.[22] When Seattleites began to reimagine their city in the 1960s, they did so within this larger national context but acted on the local and specific concerns of birdwatchers and Indians. In the national debate and on a local level in Seattle and elsewhere, the idea and reality of open space would gradually fill with different expectations for calming and even curing the various ills of an American landscape that a growing number of social critics saw as afflicted by greed, deadening rationality, and moral decay.

The growing importance of open space could be seen as early as 1958 in the *Seattle Argus*, a Seattle weekly, which called the old fort a "useless and costly military installation" and suggested that the government release the property for the world's fair planned for the early 1960s.[23] Five years later, in 1963, as Seattleites began to feel the full effects of suburban development on the edges of their city, the *Argus* made another plea, this one for using the fort as a park rather than letting it be developed, frustrating the "real estate operator's dream of tantalizing profits that could be made with those choice 640 acres." The paper argued that "civic leaders should be considering the possibility that some of that land just might become available for a magnificent park."[24] On April 24, 1964, when Secretary of Defense Robert McNamara announced that 85 percent of Fort Lawton would become surplus property, Seattleites suddenly found that they, too, had several hundred acres of this potential open space to contemplate. When the land was given to the government, it was "swampy, heavily timbered and cut up by deep ravines," according to the *Argus*, which nonetheless added that the army had "done wonders to the property" and suggested that "if it were to be de-activated no finer site could be found for a new city park."[25] Fort Lawton was only the start. By the late 1960s, as military demobilization continued and industrial plants and railroad corridors in the city fell dormant, Seattleites began to project ideas onto other such urban spaces. All these visions began to coalesce around this new concern for open space.

The Johnson administration's Park Practice Program published a magazine, *Trends in Parks and Recreation,* that exemplified the reigning opinions among public officials regarding urban aesthetics and public space. *Trends* helped to disseminate this spatial or environmental aspect of Johnson's broad expansion of liberalism during the mid-1960s. The Great Society successfully incorporated the conservation and ordering of diverse physical landscapes within this vision of the common good.[26] The inaugural issue's cover story was written by Secretary of the Interior Stewart Udall, who rallied the "taste-makers" and the "ambassadors of aesthetics, of good land use, and of national conservation attitudes" to attack blight.[27] Between 1964 and the early 1970s, *Trends'* pages described the government's new commitment to "beauty" for the lay public and armed local grassroots groups, such as the SAS, with a persuasive vocabulary and rationale for their actions. In tones similar to those of their Progressive Era counterparts, the government experts writing in this and similar publications explained that "sensory experiences," beauty and ugliness, were "directly related to the health, welfare, safety and morals of people."[28]

Trends echoed Johnson's conservation agenda, and its content inspired local groups to do everything from burying unsightly power lines and regulating billboards to creating nature reserves and vest-pocket parks for inner-city neighborhoods. The magazine also published speeches by the president on the subjects of blight and conservation, including "The White House Message on Natural Beauty," in which Johnson outlined this broad new conservation agenda. Well before Earth Day and the use of *ecology* in the common parlance, his speech gave wide currency to some of the main directions in which the environmental movement would evolve over the course of the late 1960s and 1970s. This "new conservation," Johnson explained, "must not only protect the countryside and save it from destruction" but also "restore what has been destroyed and salvage the beauty and charm of our cities." Unlike the "classic conservation of protection and development," the new conservation, he said, would be a "creative conservation of restoration and innovation. Its concern is not with nature alone, but with the total relation between man and the world around him." In accordance with the general thrust of the Great Society programs that addressed urban poverty and health, Johnson's speech outlined environmental remedies that always began with the city, including creation of the "Open Space Land Program,"[29] which aimed at "improving the natural beauty of urban open space," along with other federal efforts to protect water and land. In the nascent ideology of postwar environmentalism, Johnson placed cities and people at the center of his environmental vision.[30]

Enhancements of beauty, according to Johnson, were the "principal responsibility of local government, supported by active and concerned citizens."[31] Such sentiments appealed to a variety of would-be activists in Seattle. The president invited everyday Americans to attack blight and ugliness, but he naturally called first on the experts: professional planners, parks directors, and scientists. *Trends* provided news and the opinions of these experts and policy makers.

The same language of blight and beautification that downtown developers would use to justify their transformation of downtown could just as easily have applied to plans at Fort Lawton. *Trends* demonstrated the potential influence of geographers, forest and wildlife managers, and especially landscape architects on the early formation of 1960s environmental attitudes in the city. These experts provided the 1960s equivalent of the City Beautiful movement's proliferation of expertise and zeal for problem solving.

Unlike their Progressive Era counterparts, however, the Great Society experts developed a vision of beauty in the city that suggested a stronger critique of capitalism, a critique foreshadowing the rhetoric and attitudes that would animate activists in the following decade. In "Conservation and the City," for instance, the geographer Donald Goldman argued that "the social and emotional ills facing the city" displayed "disturbing evidence that perhaps human well-being is not solely dependent upon speed or efficiency or economics, that perhaps we also require beauty and stillness, considerations that have often been ignored because they did not come under the heading of 'progress' and could not be valued in dollars." Looking for "ways to reestablish some contact with the land and to bring more beauty into our cities," Goldman argued that the collective effort facing America was "not merely a back-to-nature fad, but a sincere effort to achieve a better form of city and a better way of life by providing for human esthetic needs."[32] In a 1966 *Trends* article titled ". . . On Beautification," the landscape architect Richard Moore echoed Goldman's concerns about capitalism, efficiency, and the well-being of the city-dweller deprived of beauty. "Sensory experiences" of beauty or ugliness, argued Moore, are "directly related to the health, welfare, safety and morals of people." He added that "we cannot have beautiful urban centers if we and our representatives allow greedy, narrow minded and short-sighted merchants to dictate and create our urban environment, uninformed, and uncontrolled."[33]

According to the experts and government officials writing in *Trends,* the "beauty and stillness" that nature offered to a postindustrial city in an affluent society formed a crucial aspect of an emerging urbanism. But creating a controlled, ordered, and healthy urban landscape required both a chastening of capitalist excess and the proper guidance and information that they could supply. They thus encouraged "formal instruction in environmental appreciation" at all levels, from elementary school to adult voters.[34] A "symptom of a sick society," the ugliness of the city could be eliminated if an "enlightened populace, enlightened representatives, and highly qualified specialists and designers" acted.[35] With the president's encouragement and power behind them, they spread the word, and like their City Beautiful predecessors, they hoped to boost their own professional standing at the same time.[36] Although *Trends* emphasized nature as a crucial and calming component in a tumultuous mid-1960s city, such ideas also echoed the efforts of urban experts such as Victor Steinbrueck and the other market preservation activists. But this language was up for grabs and could be used and interpreted in contrary ways.

Within this idea of open space and new parks for urban America grew the possibility for reimagining large sections of urban landscapes. Control, order, aesthetics, wilderness, leisure, and "the good life"—however these ideas and oft-used phrases were understood by experts and everyday Seattleites, their vocabulary played a part in the open-ended conception of urban planning by middecade. In this moment before the popular ecological turn in 1970, the government and conservation enthusiasts had begun to unleash the unpredictable power and varied agendas of both grassroots and elite activism. The shape of the city and nature within it became the subject of often competing visions of the city. Beauty, blight, and open space were part of the lexicon for all who hoped to influence the taste and aesthetics of their communities or simply to control the production of urban space. In the rapidly shifting political and economic conditions of the postwar city, however, nobody completely controlled the use of this rhetoric. The civic-minded discourse of the Kennedy and Johnson era gripped Seattle in the 1960s, and like the urban preservation movement, it inspired a variety of actors to think about the city's future and past and about the role nature could play in its changes.

By middecade local tastemakers—including downtown businesspeople, preservationists, and Model Cities activists—couched urban poverty, disorder, or blight in these environmental terms, but it remained unclear to whom this language would give power and what or who would be the object of that power. Poverty and the "urban crisis" would increasingly preoccupy urban liberals, but rather than aggressively address the roots of urban inequality—as was done in the early War on Poverty—critiques turning on these notions often overemphasized conditions of the urban environment itself as a cause for poverty and other social problems.[37] The underlying inequalities, such as discrimination in housing, lending, and workplaces, that produced these problems were either harder to address or simply ignored, as they were by the many urban leaders and voters who

United Indians of All Tribes demonstrating at Fort Lawton, 1970. *Seattle Post-Intelligencer* Collection, image no. 1986.5.51939.1. Courtesy Museum of History and Industry.

had opposed open housing and other desegregation efforts earlier in the decade. Both local and national leaders instead saw their mission as improving the city's appearance and character, and usually this emphasis benefited middle-class whites. The shortcomings of these ideas emerged plainly in the battle of Fort Lawton.

Dueling Ecologies

As diverse constituencies staked their claims to urban public spaces and public speech during the 1960s, the rhetoric of open space became more urgent and more complicated, especially viewed against the backdrop of urban violence and disorder. Near the end of the decade political divisions over civil rights, the Vietnam War, and gender equality crystallized in emblematic confrontations and violence in Chicago, Berkeley, and elsewhere. Seattle's public spaces were similarly fraught with social and political tensions. Narratives about the meaning of "open space" or environment would proliferate as a variety of actors sought political power in the city streets and public places.[38] The idea of open space at the fort may have emerged in an early 1960s context of beautification, an empowering aesthetic message, but by decade's end it would be clear that even seemingly benign arguments about aesthetics involved political authority as well. On April 24, 1964, when Secretary of Defense Robert McNamara announced that 85 percent of Fort Lawton would become surplus property, the fort emerged as a green tabula rasa on which the city's political conflicts and hopes would be written.

The women active in the local chapter of the Audubon Society were the first and most important actors to reimagine the fort.[39] Although these women formed a crucial part of the larger coalition fighting for open space, their interests, unlike the varied agendas of other groups in the coalition, focused specifically on using (and helping to manage) the fort as a wilderness habitat for birds. While other factions in the CFLP envisioned everything from a golf course or rhododendron gardens to a more traditional park, the women of the SAS crafted a comprehensive vision from the start.

In no way were white, middle-class women and Indians' historical oppressions equivalent, yet the strategies of both the SAS women and the UIAT showed how the politics of identity and personal power, whether for women or for people of color, were inseparable from the green politics of the late 1960s. Both sets of actors showed how the word *ecology* in popular use provided a new authority and utility while revealing the political (and personal) nature of any claims made on urban open space. In this way the story of the UIAT and the SAS women—although divided by class, ethnicity, and historical experience—shared a political context in which the hegemonic claims of postwar liberalism had begun to erode. Both groups were instrumental, wittingly and unwittingly, to reshaping environmental discourse from a unified story of Great Society beautification to a more complicated and contested story that featured ecology and contests over public space. Although each group of actors was on the fringes of power, urban environmentalism provided both an entrée to political engagement.

The local chapter of the SAS saw its membership grow exponentially in the second half of the 1960s, and its mission also changed to become more obviously civically engaged. In particular, the group provided middle-class women another context for asserting political influence through advocacy and education in the public realm. Bird-watching gave women license to survey potential open spaces in the city and to imagine new kinds of public space.[40] Women had derived authority from their traditional gendered role as "urban homemakers" in the past, but the SAS increasingly employed the language of popular ecology in the wake of Rachel Carson's *Silent Spring*. Stephen Bocking, a historian of ecology and its political and institutional contexts, distinguishes between ecology as "a source of moral values or a guide for those wishing to live gently on the earth" and the scientific discipline of ecology, which has "its own priorities and history."[41] Groups such as the SAS operated in the former mode, using ecology as a moral guide and authority in their efforts.

Unlike the women who challenged freeways earlier in the 1960s, who focused on aesthetics, the SAS applied this often loosely defined but effective political discourse of ecology to the debate over public green space. Terms such as *nature, ecology,* and *open space* blended together in SAS rhetoric, though the women still emphasized familiar themes of social control. Seattle Audubon Society activists argued, as had their Progressive Era counterparts, that green space for bird habitat was necessary to bring moral order and edification to a restless city.

Indeed, during the late 1960s SAS newsletters show the club turning its binoculars increasingly on the city itself. The longtime secretary of the SAS, Hazel Wolf, and other SAS activists emphasized acquisition of open space. The group's members had previously concentrated on weekend field trips to out-of-city bird sanctuaries and nature areas, presenting nature films and slide shows in church basements, and other expected activities. When they went bird-watching within the city, they spotted birds in a few predictable places, such as Seward Park or Golden Gardens (a beach on Puget Sound). They dutifully counted the birds and reported their findings in the newsletter. Every Christmas the society performed its annual bird count, a process that required members to stalk around in places not always associated with nature: under bridges, near factories, and in empty lots. Soon, however, the club would begin staking claims to habitats in often unexpected places. Fort Lawton topped the list. In 1964, as Seattleites heard the news that the land would be made available, Irene Urquart announced a field trip to the army site. Expect "migrating and resident Warblers, Vireos, Sparrows and animals of the area," she wrote. Coming just months after the Department of Defense announcement, the trip would be, she said, "an opportunity to see what we are fighting for when the Fort is de-activated."[42] City and federal authorities and developers were eyeing the prime real estate on the bluff, but Wolf, Urquart, and other SAS members had already done so. They called on their members to join "the fight."

That year the SAS developed a more activist orientation, publishing more strident articles about habitat preservation and urban park recreation. The local chapter also published Johnson's speech on natural beauty in its April 1965 newsletter.[43] In May one SAS member, Theron Strange, wrote an energetic feature after attending the Pacific Northwest Recreation and Parks Conference, a meeting whose tone closely reflected the editorial opinion of *Trends*. Strange reported that more than "400 professionals and parks recreation people" attended the conference. Presentations at the conference echoed Great Society beautification rhetoric, emphasizing the need for recreation in an affluent society. One speaker argued that American culture should be reshaped from an older survival mode to an "affluent society in which leisure is a controlling element." And he claimed, "We need to develop individuals in a planning society, not pawns in a planned society."[44] Strange reported that one speaker at the conference even drew inspiration and tactics directly from the civil rights movement's mobilization of young people, arguing that new conservationists would need similarly "to risk sanctions" and "plunge into politics" in order "to gain recreational goals."[45]

By 1968, with political change emerging throughout the nation, the SAS and its growing membership reflected the activist spirit in full bloom. The newsletter detailed ongoing fights to "save" various urban landscapes that the SAS found critical for birds. By decade's end, the group scheduled more regular visits to urban parks and bird-watching sites that were not yet open to the public, including Fort Lawton, the soon-to-be-decommissioned Sand Point naval air station on Lake Washington, and the Union Bay marsh area near the University of Washington.[46] One member reported on the effort to preserve the "wild marsh," arguing that although such places "are miniscule," they nevertheless "are extremely important for reasons of wildlife welfare, for ecological and aesthetic values for study and scientific work, and even for limited recreational use."[47] Thanks in part to the SAS's advocacy for bird territory, Seattleites began to agitate for open spaces that would provide more than pleasant views.

Whereas the women of the SAS subtly enhanced their political power with environmental activism, the Indians who invaded Fort Lawton were blunt. Two important and specific political contexts framed their claims.

First were the "fish-in" protests in Washington State during the early 1960s. In 1964, a little more than three months before the women of the SAS celebrated the announcement of surplus land at Fort Lawton, Native Americans confronted federal officials along the rivers in southwestern Washington. Tribal leaders, including many women, battled for their rights to fish for salmon in the state's rivers after state officials began a drive in 1961 to end tribal fishing rights.[48] In their fight to maintain treaty rights to fish in "usual and accustomed places and stations," Indians in Washington found themselves facing an array of forces including sport anglers and especially state enforcement officers, with whom the tribal groups had a series of well-publicized standoffs. These "fish-ins" were modeled after the civil rights "sit-ins" and drew widespread public attention as Indians fought—eventually with success—for their right to fish on state rivers.[49]

The fish-ins, then, constituted one context informing the story of contested urban public space and the Indians' actions at Fort Lawton. The protests pushed forward the language of ecology, though highly contested, as a powerful tool in 1960s political discourse in Washington State. As Indians asserted their treaty rights, the ensuing conflicts pitted western Washington tribes against outdoor groups who claimed to be conservationists as well. Indians faced blatant discrimination and racism from state officials, wildlife and sport fishing supporters, and others who feared that they would deplete a finite resource. Only later, in the 1970s, would Indians and conservationist groups, facilitated in part by women such as Hazel Wolf of the SAS, begin to make important coalitions around environmental issues.[50] The fish-ins, which drew Dick Gregory and other famous civil rights activists to Washington State in 1966, also linked what appeared to be a local protest to the national political culture of the civil rights era. These links became more apparent as the mostly young activists who would later demonstrate at Fort Lawton, empowered by the fish-ins locally and by the Red Power movement nationally, began to radically claim urban places. The Fort Lawton leader Bernie Whitebear—a participant in the fish-ins, a Seattle community activist, and yet not a member of a western Washington tribe—exemplified the convergence of these political threads in Seattle.[51]

The urban experience of young Native Americans during the late 1950s and early 1960s formed the second broad context that helps to explain the assertion of Indian claims to Fort Lawton. According to the historian Coll Thrush, historical treatments, as well as stereotypical images of Native Americans, have helped to perpetuate the notion that "Indians and cities are mutually exclusive."[52] Groups such as the UIAT actively confronted such notions. Indian hopes for the fort spoke directly to an experience of urban erasure and ghettoization—and by extension, the shared experience of exclusion among people of color—in West Coast cities at the time. From 1950 to 1970, Seattle, Portland, San Francisco, and other cities in the West acted as relocation centers for young Native Americans, a result of the federal government's policy of moving tribal peoples off rural reservations and into the "mainstream" of urban life. The Bureau of Indian Affairs moved Indians to urban areas near industry, promising training in welding, appliance repair, and other trades. Mostly unsuccessful as an employment project, relocation often left undereducated and unprepared Indians to fend for themselves in urban ghettos and to deal with widespread job discrimination.[53] As John Trudell, a prominent Red Power activist who shared this early 1960s experience, said during the Alcatraz occupation, "Coming into the city as a stranger is probably one of the worst things that could happen. Some day all of that is going to change."[54] His statement suggests the sense of exclusion and dislocation that these urbanized Native Americans felt acutely.

Although Indians have always lived in Seattle and helped to shape its history, in the 1950s and 1960s they also shaped stronger urban communities in response to the shared experience of racism and displacement there. In larger relocation centers, such as San Francisco and Seattle, according to one observer who moved between these communities, Indians created "their own little Indian reservations in the city."[55] In Seattle, for instance, a group of Indian women started the Seattle Indian Center in 1958. Later, during the late 1960s, this center disseminated information about the protest at Fort Lawton in its *Indian Center News*.[56] Though often woefully underfunded and burdened by community problems of alcoholism and poverty, these Indian places were the nuclei of the movement. Indian ghettos were rife with despair, but they also created the conditions for a diverse and multitribal base that would help to build Indian political power in Seattle and other cities. Around the time of the Fort Lawton occupation, Indian centers—meeting places funded by community action programs in Seattle, Portland, and San Francisco—functioned as loci of organization

and activity for an increasingly assertive Indian youth movement that began to challenge the more incremental politics of the time. Thus the historian Nicolas Rosenthal argues that Portland Indian activists at this time
were part of a larger shift in urban Indian political representation that increasingly emphasized the identity-based radical politics of the larger
youth movement.[57] As did their counterparts in Portland, the Seattle activists represented a shift in power from the more staid urban organizations
(as well as the conservative, rural-based tribal groups involved in the fish-
ins) to the more forceful politics at Fort Lawton. The Fort Lawton occupation signified a symbolic and actual reclamation of land (though the claim
was never based on an existing treaty dispute), but perhaps more important, it announced a new political style and aggressive presence of margin-
alized actors in the political life of a city that hitherto had ignored them.[58]

Two years before the Indians would first invade, on March 15, 1968, after tantalizing Seattle with visions of open space, the Defense Department
withdrew the offer of surplus land and set in motion the events that would
follow. The government explained that the land would likely be necessary
for military use as reserve troops were called up for the Vietnam War. The
government also dropped a bombshell: not only would there be no park
or urban "wilderness" for birds and bird-watchers, but the government
planned to use Fort Lawton as an ABM missile base. With a launch site
from the Puget Sound area, the government argued, the new Sentinel missile system could effectively defend the United States from Soviet and Chinese missiles. In an instant the open space that SAS members had walked
and surveyed, that they had assumed would be made available to them
by existing surplus property laws, was gone. The war machine was coming to their imagined garden in the city.[59] Both the women of the SAS and
the young Indians of the UIAT watched as the first battle for Fort Lawton
began.

The Battle for Fort Lawton

With the threat of urban-based missiles to galvanize them, a cross section
of mainstream civic groups, especially environmental ones, gathered to
challenge the government's plans. By summer of 1968, the group calling
itself Citizens for Fort Lawton Park began to rally diverse groups of activists, including the Sierra Club, the SAS, and civic groups such as the Junior
League. They sought to challenge the arrival of missiles in their urban wil-

derness. One historian argues that this coalition was essentially conservative, driven by the attitudes of new conservation rather than any Vietnam War–era antimilitarism. The hikers, birders, and neighbors in this broader coalition were motivated first by a vision of open space that everyone could enjoy, but many also thought that such places would serve as an outlet for the potentially destructive energies and anger of urban youth. Seattle elites in the past had enticed the armed forces with gifts of land and were denied. By the Vietnam War era, when the government was finally ready to bring its massive firepower to the city, this coalition wanted a park and the domestic tranquility it might provide.[60]

The first skirmish in the battle over Fort Lawton then began. An attorney named Donald Voorhees spearheaded efforts to organize various civic organizations. He used the fort's history—especially the fact that Seattle had donated it in the first place—as a way to convince the federal government to abandon its plans and to arouse public interest. In a lengthy article in the *Seattle Post-Intelligencer* during October 1968, Voorhees asked simply, "Will Lawton become a missile base or a park?"[61] The paper reported that the battle would be "possibly coming to a head in the next few weeks" and that the fort controversy might be "a milestone in that it may show citizens in the Age of Bureaucracy still can be heard before the moment of decision-making at a high level in Washington, D.C." There seemed to be "no disputing," one reporter announced, "that a loosely knit group of citizens and local politicians" had managed to move "the immovable Pentagon Brass," and the journalist pointed out that it was being done "with factual arguments rather than political horse trading."[62]

In reality, however, by the time the fight had gone public, the women active in the SAS, along with the Seattle mayor's office, had already held discussions with Washington's U.S. Senate delegation and the Defense Department. The SAS was decisive to the outcome. Unlike other coalition members, such as the Junior League or the League of Women Voters, the SAS had distinct plans for Fort Lawton. After one CFLP meeting held on the university campus, the SAS newsletter reported, "We have a more than ordinary interest in the success of this organization. . . . Think of the bird walks we could have so close to the heart of the city!"[63] In subsequent issues, the group's newsletter encouraged members to write their representatives and Senator Henry M. Jackson, who were "working hard for this park."[64]

119

In early August 1968, before Voorhees assumed a leading public role, SAS members sent a letter to Lieutenant General Alfred D. Starbird, a system manager for the ABM program, in which they expressed their "position," which, they emphasized, their organization had "stated publicly four years" earlier, following the original announcement of the government's intention to make the land surplus. Using familiar Great Society beautification rhetoric regarding open space, they pointed to the growth of urban centers, the ratio of parks and recreation space to population, and general congestion as reasons for their continued "resolve" to make the fort into a park. They argued that long-standing multiple-use policies for public land ought to be amended to include present-day needs, including "spiritual, esthetic and recreational considerations." Members described their own survey of the land in 1964, which had revealed many uses of the park for picnics, sports, and other organized events. "But what most delighted our surveyors," they told the lieutenant general, "was the unspoiled wildness of much of the area," with the "air filled with birdsongs of many species, some of them rarely seen within the city." They reported that "small mammals lived in the woods," as did a variety of trees, wildflowers, ferns, and shrubs, and that they had "discovered a year-around stream and rare marshes." The SAS hoped to convince the government of the necessity for urban wilderness.[65] Yet while the arguments for beautification had morphed into an argument for a wilderness park and bird habitat, the familiar concern for order persisted within this emerging rhetoric of popular ecology. "Open spaces within and around cities," they argued, would be "important elements in any plan to alleviate the conditions in our cities that lead to poverty, crime, and civic disorders."[66] In the era of street protest and fears of riots and unrest, the SAS members thought nature could serve to regulate the city, providing, as the writers in *Trends* had argued, for the "welfare, safety, and morals of people" there.

Ironically, the SAS and CFLP would mimic the street fighters of the late 1960s with their own version of post-1968 protest action—including pickets and even a subtle occupation of the fort—to get their wilderness. In late November 1968, some three hundred Seattle residents converged on the fort to demonstrate for a park. As the neighborhood's *Magnolia Times* put it, the Seattle Audubon Society and other groups that composed the Citizens for Fort Lawton Park made a showing "in another skirmish of the 'battle for Ft. Lawton.'" According to the paper, "they were met at the South Gate" by the post commander and military police, who allowed them in but "required them to leave their placards outside the gate."[67]

Politicians paid close attention and began to respond to this outpouring of interest and grassroots activism on behalf of open space. Senator Henry M. Jackson was one of those politicians. The Citizens for Fort Lawton Park and its component groups had long pressed Jackson for help. Well known for his efforts to preserve wilderness areas in Washington State, he would use his influence on powerful committees as political weight, along with personal persuasion, to fight similar battles for cities. His efforts on behalf of Seattle's urban green spaces, beginning with his support for the Fort Lawton cause, would have profound repercussions for open space throughout the country. Thanks in no small part to Jackson during December 1968, the Pentagon decided to back down, announcing that the ABM base would be relocated. In March 1969, however, when the *Seattle Post-Intelligencer* headline read "Seattle May Finally Acquire Ft. Lawton as Park Site," neither the fate nor the meaning of the open space had been settled.[68]

That March Jackson outlined legislation that would allow cities to more easily obtain surplus federal property, and his description of wilderness again matched the social control agenda of Great Society beautification rhetoric. An amended version of the earlier 1965 Land and Water Conservation Act, the new act would be labeled the "Federal Lands for Parks and Recreation Act of 1969." Before an audience of planners in Seattle that year, Jackson explained, "Since 1964, my Committee has approved and the Congress has enacted legislation to set aside 13 million acres of land and water for the use and enjoyment of the American people." Yet, he said, "I am not optimistic that the Nation is winning, or will win, the war for 'inner space'; the battle to have an ordered environment in which society can achieve its aspirations." He explained that "improvement of the environment . . . can mean most to those who are most deprived." Jackson underlined nature as the force for alleviating the disordered human and physical environments of the nation's cities: "Because I believe in the relevance of parks and recreation programs to our urban crisis," he told the assembled planners, "I recently sponsored the 'Federal Lands for Parks and Recreation Act of 1969.'" Pointing to the efficacy of urban parks, he said, "I think there is a definite connection between congestion and violence."[69] Again, the reasons for creating these nature spaces were numerous, but the fear of urban disorder remained central to the rationale of wilderness advocates and bird-watchers alike.

In a fractious time, such rhetoric had significant influence. According to the amended act, the secretary of the interior was empowered to dispose of "surplus real property, including buildings, fixtures, and equipment, . . . as needed for use as a public park or recreation area."[70] The consequences of the act rippled across the country, making it possible for any American municipality to claim surplus property for free. Before it was passed, cities either would pay market prices for the land or it would enter private hands. Jackson's Fort Lawton bill (i.e., the Federal Lands for Parks and Recreation Act) paralleled his more comprehensive landmark legislation the same year, "The National Environmental Policy Act of 1969."[71] That act contained familiar rhetoric and ideology rooted in the Great Society agenda, which sought to beautify, preserve, and restore the American landscape. In this effort to save cities from blight, liberals sought a balance between "man and his environment" and emphasized "productive and enjoyable" experiences of nature. The environmental policy act also expressed the evolving concerns of groups, such as the SAS, who used the language of "ecological systems." And significantly, it reinforced their call for enrichment of "understanding," an invocation to use public space and park lands specifically as educational forums about natural systems.[72]

While the Senate debated Jackson's legislation, SAS activists Hazel Wolf and Ann Mack had already begun to remake the fort as a wildlife preserve that could teach urbanites about ecological systems and the civilizing effects of nature. The Seattle Audubon Society shaped the park's outcome in part because it could use the connections and resources of a powerful national activist group. In fact, the SAS's Nature Center Planning Division encouraged Wolf and Mack to create an Audubon Society nature center at Fort Lawton. The director provided steps that Wolf and Mack might follow in "establishing a nature center," including cultivating "local sentiment."[73] While the precise use of Fort Lawton was still undecided up through early 1970, the local Audubon Society chapter proceeded quickly. Before the city or any other entity had completed a survey or plan, the SAS was already at work on *A Fort Lawton Park Nature Center: A Preliminary Proposal*, published later in 1971.[74] The flood of environmental legislation that emerged from this late-1960s context therefore reflected, in part, the local class interests and concerted efforts of middle-class activists who acted within the intense atmosphere of urban politics and urban fears to find environmental solutions.

Right of Discovery

On March 9, 1970, when the United Indians of All Tribes arrived on the bluffs of Fort Lawton, they took the SAS's Great Society ideas about urban nature and turned them upside down for their own political purposes. Unlike their Great Society counterparts, or the mainstream civil rights organizers in the Model Cities program, the people who occupied Fort Lawton were more radical. Just as the predominantly white, middle-class birders and beautifiers were convincing the government with their arguments, the UIAT arrived with dramatic flourish to define their own ideas about open space and nature in the city. The UIAT hoped to protest in much the same way as had the CFLP a year earlier, marching into the fort and symbolically claiming land. They met a far less friendly reception. Bob Setiacum (Puyallup), a "fish-in" protest leader, read the Indians' proclamation of discovery; as he did so, the *Seattle Post-Intelligencer* reported, his words were "drowned out because Army MPs began carting people away." Unlike the SAS or Sierra Club protesters, the UIAT complained that MPs purposely dragged them through sticker bushes and nettles during their "invasion." Seattle papers played up what they saw only as an absurd spectacle with playful and casually racist headlines such as "Indians Drum Up Support for Fort Claim" or "MP Now Knows 'How Custer Felt': He Shouted 'Stop' But Indians Swarmed Into Fort." Journalists emphasized the drama of the invasion and subsequent occupations that early spring.[75]

Images of MPs in full riot gear at Fort Lawton followed closely on the heels of a similar Indian occupation of Alcatraz Island in San Francisco and the infamous fight over People's Park in Berkeley in 1969. Activists from Alcatraz and other participants from the Bay Area's cross-fertilizing protest culture—including Russell Means and Grace Thorpe—brought their media savvy to the cause in Seattle.[76] The overreaction of fort police focused even more attention, while Voorhees and the CFLP expressed worries in the *Seattle Post-Intelligencer* that the Indians' protest might short-circuit their own claims.[77] But the Indians also brought a new set of political energies and orientations to the table. The Black Panther Party and the Seattle Liberation Front, for instance, both offered "psychological clout" to the Indians, according to the *Post-Intelligencer*. Many young white activists, a counterculture that often glorified stereotypes of the "ecological Indian," supported Indian protests during this time.[78] "Uncivil acts"—urban unrest and the spectacle of the antiwar movement—had arrived at the gates of the very open space that middle-class activists in the coalition had hoped might calm such radical displays among Seattle's urban youth.

The Native American activists who invaded Fort Lawton may have emerged from the same cultural ether as their beautification counterparts, but they introduced a very different perspective on the uses and meaning of public space and nature. White liberals hoped to balance the good life with the changing metropolitan region. They were willing to fight downtown business leaders, urban renewal, and even freeways while advocating open space in the process. But the Indians at Fort Lawton forcefully articulated the concerns of groups who had been most excluded from this metropolitan project in depressed areas such as the Central District, Pioneer Square, and the International District. Those with the least power in shaping public space or defining didactic functions for urban green spaces began to make their voices heard at Fort Lawton.[79]

In an especially ironic twist, the UIAT inverted the SAS's Great Society didactic rhetoric about alleviating urban dysfunction though open space; according to the *Seattle Post-Intelligencer,* the UIAT wanted to use Fort Lawton first of all to "set up an 'environmental preserve'" that would "help teach whites 'how to stop destroying the earth.'" Like the SAS, Indian activists had their own idea of open space as instructive: "most of the land would be restored to its most supportive natural state," they argued; "There would be berries, trees, game."[80] As did wilderness park advocates, the UIAT used wild nature and the language of popular ecology in their plans for the proper use of the fort. They also manipulated the stereotype of the "ecological Indian" to claim authority over the fort, arguing that Indians held special "lore" regarding the land and how to "save it from destruction." Most of all, the UIAT claimed territory for reasons similar to those given by the women of SAS: they presumed to address, by providing open space and their own education programs, "a base from which to operate," according to Whitebear.[81]

The Native Americans at Fort Lawton combined contemporary ecological concerns with the already firm stereotype of Indians as stewards of the land. To justify their action, they adapted this stereotype, blending it with

Bernie Whitebear, 1971. *Seattle Post-Intelligencer* Collection, image no. 1986.5.55140.1. Courtesy Museum of History and Industry.

the sort of language of ecology that the SAS employed. Stereotypical images of Indians already permeated the countercultural and environmental movement by the late 1960s. Indeed, appropriated Native American imagery was central to many countercultural claims. Antiwar and counterculture activists often masqueraded in Native American costume as part of their oppositional politics.[82] Hippies and New Left activists wore stereotypical Indian garb of headbands and beads. They exploited Indian-related imagery to make arguments about everything from environmental degradation to occupations at People's Park and elsewhere. The historian Philip Deloria argues that throughout U.S. history—from the Boston Tea Party to the present—white people have used the rhetorical power and associations of "Indianess" to negotiate American identity in moments of historical change or disorientation. Deloria maintains that this use of Native American imagery in the protests of the 1960s was especially troubling because of a tendency to detach symbols from their meanings, with historical memory, the specificity of past injustices, easily lost in the process. If anyone could be Indian in the free flow of cultural referents, De-

125

loria asks, if whites could deploy "Indianess" and historical associations of ecology and governmental oppression on behalf of their own liberation struggles, then what did it mean to be a real Indian with a real history of oppression?[83] Yet the pan-Indian activists and local Indians at Fort Lawton claimed a no less dubious position of authenticity and moral authority when they made their arguments about their unique authority as environmentalists who could restore the land to its "most supportive natural state." In these ways, then, Fort Lawton activists drew from recent experience and from stereotypes of themselves as ecologically sensitive stewards of the land. In the context of 1970s Earth Day politics, this moral authority regarding "defiled" urban landscapes provided persuasive weight to their claims, worrying the CFLP further.

Ecology took on a different meaning in this context. Unlike the Great Society, with its preoccupations with order and aesthetics, the UIAT did not envision a "naturalized" city. Moreover, unlike the SAS's popular notions of ecology—which represented a maintenance of "wild" open space habitat for birds (and less overtly, for birders)—from the beginning the Indian vision emphasized *use* and a mingling of social need, cultural markers, and nature. Despite their ecological rhetoric, the UIAT's elaborate plans for Fort Lawton involved much more than communing with or protecting nature. Theirs was a different kind of symbolic and practical landscape, a landscape that could combine responses to urgent social and political needs with a concern for nature. Although similar on the surface, this redefinition of open space differed markedly from those of the SAS and other civic groups or Senator Jackson's ideas about wilderness as a way to channel disorder from the city itself.

The Indian pickets and their supporters continued outside the fort gates throughout late March 1970, their claims joining with these other ideas.[84] In the wake of these events and in preparation for Senator Jackson's impending visit to the fort in the spring of 1970, Donald Voorhees sent a letter to the lawmaker offering talking points he might use to address the new roadblock in the park advocates' plans. Addressing neighbors' fears about urban Indians, Voorhees wrote, "Residents are concerned about the creation of a large public park. . . . They fear that it will be flooded with hordes of uncontrollable persons. They would really prefer to have Fort Lawton remain just as it is." With Native Americans now in the mix and their conspicuous presence in the affluent neighborhood, area residents became apprehensive about open space. But Voorhees dismissed such fears as groundless: "There is no reason why a park at Fort Lawton should be less safe, less peaceful or less orderly than any other major park in the Seattle park system." Responding to these fears, Voorhees used an even older rhetoric of nature that Native Americans might have found all too familiar. "These Lawton lands will do the greatest good for the greatest number," he wrote, "if they are developed as a whole for the whole community" instead of being "parceled out to various groups" or designated for the "enjoyment of only one segment of the community: those of Indian extraction."[85]

After months of negotiation, the city and the UIAT came to an agreement in which the group would obtain a lease to 17 acres of the fort's lands for a cultural center. Richard Nixon officially transferred the 391-acre site to Seattle on September 1, 1972. With help from its national office, the SAS pushed for and got its long hoped for "community nature center." Explaining its planned use of the land, the SAS described a future park that would incorporate what it saw as the didactic, civic, and even patriotic qualities of nature. This new nature center would combine the previous era's concern for beautification and order with the post–Earth Day language of "ecology." Nature at Fort Lawton would bolster democratic ideals and citizenship while allowing visitors to find humanity's "place in the ecological community," the group argued.[86]

Soon after it took possession of Fort Lawton, the city commissioned a master plan for the new park that would elaborate on the SAS's original hopes for the space. The first revised plan, appearing in 1974, featured the park's new designation: "Discovery Park." First pronounced ironically by Indian occupiers, planners reclaimed the language of discovery in honor of Captain George Vancouver's ship, which had plied the waters of Puget Sound in the late 1700s. Discovery, science, and proper nature appreciation dominated the narrative of the park experience in just the way the SAS and wilderness advocates had hoped. Dan Kiley, the author of the master plan, described the novel focus for the new wilderness parks: "educational features" that he said "should not dominate but become an integral part—just as sculpture and small pavilions have enhanced parks traditionally." Environmental and cultural values were to be inscribed in the park's landscape. Indeed, in the years since its origin, the city has defined and actively managed the park as a "wilderness," restricting certain uses and taking great pains to remove nonnative plant species. Kiley also hoped that operations of park programs could be shared by "Environmental Organizations, Nature Societies, Recreation and Outdoor Clubs, etc." The training that these groups provided would, it was hoped, inculcate the proper appreciation of hiking and birding, and nature-oriented groups, such as the SAS, would have a place in influencing the training and the experience. The cover photograph of the revised Discovery Park plan shows a group of young people and adults, circa 1974, inspecting leaves and enjoying the view, while to the left of the group stands a uniformed park ranger. Here the roles of authority and order are revealed, fitting together with nature appreciation.[87]

Yet at Fort Lawton today Native Americans continue to control an autonomous place. Still operated by the UIAT, the Daybreak Star cultural center, born out of the 1970s activism, continues to run a Head Start program for urban Indian children and to sponsor Indian art displays and cultural events.[88] Bernie Whitebear directed the cultural center until his death in 2000. His memorial service, held in a massive hall at the Washington State Convention Center, drew a diverse group of mourners, including U.S. senators, union activists, and 1960s activists of many ethnicities and political backgrounds. Just as Whitebear had argued, Daybreak Star became a place where Indians could simply exist on their own terms, but more important, a place from which urban Indians could organize, politically and culturally, close to a natural setting. The UIAT did not obtain the entire open space, but through their negotiations they obtained a piece of land that sits comfortably next to a significant tract of parkland near the urban core of Seattle. Significantly, the Indian culture center at the old fort does not downplay history in the landscape, as the wilderness park does, but rather found a way to combine the political and cultural needs of urban and exurban Indians in Seattle. History, politics, identity, and nature cohere in one place.

The symbolic and physical triumph of the urban wilderness aesthetic in Seattle was never complete. Tensions between wilderness and history persisted. By the early 1980s, with the city beginning to repopulate after the economic downturn of the 1970s, news headlines announced "The Battle of Discovery Park." In 1983 the *Seattle Weekly* reported that the latest battle was "between nature lovers and historic preservation."[89] This battle revolved around a dispute between groups of the original supporters of the park concept. Allied Arts and the local American Institute of Architects, among others who had helped to block the ABM base and build a park, had begun to challenge the wilderness-in-the-city vision of the SAS and hikers by the 1980s. Preservation advocates—now seeking to preserve human history and culture—challenged the city's plan to demolish historic buildings, fearing an erasure of human history at the fort. According to the parks department, the sites where the soon to be demolished buildings stood would be "re-landscaped and seeded with wild grasses to blend in with the surrounding natural meadows, and a few trees would be added."[90] A parks planner maintained that the point of the park had always been to "downplay human development in favor of the natural environs."[91] This effort to tear down and reseed the "natural environs" struck some historic preservationists as absurd and antiurban. Art Skolnick, a preservationist who was integral to saving the city's Pioneer Square district in the early 1970s, described such plans as "elitist." "The area," he said, "is not a wilderness; it's an 'urban facility,'" and he believed limiting the number of visitors by limiting buildings verged on being racist and "un-American." Perhaps as public space again seemed threatened in an era of increasingly privatized space, people found more solace in wilderness than in a mixed landscape of history and nature. Or perhaps, as Skolnick argued, the north-end neighborhood residents had chosen a kind of private, peaceful escape in their own backyard, a darker side of an environmentalism that departed from a potentially more collective, if contested, politics of place in the city.

The United Indians of All Tribes pragmatically grabbed hold of the ideas of ecology and beautification. In the process they helped to redefine the priorities of the environmental movement. In such actions can be found the roots of the environmental justice movement that would rise by the 1980s. Mainstream environmentalists may not have recognized the Indians' desire for space at the fort as an environmentalist claim at the time, but the occupation helped to expand the very idea of what and who constituted the popular environmental movement. Just as the advocates of preserving the Pike Place Market had begun to expand their rhetoric to include social justice as well as aesthetics by the late 1960s, the Indians at Fort Lawton used the language of ecology to draw public attention to the living conditions of Indians and poor people and their environments—specifically, the neglected neighborhoods of the city.

The United Indians of All Tribes, the group that claimed Fort Lawton in the early 1970s and hoped to teach white people to take care of the earth—though playing on stereotypical notions of Indians as conservationists—may have presented a lesson after all. The Indian culture center at the old fort found a way to combine, as Model Cities workers had hoped to do, the political, cultural, and environmental health needs of people. But the park's new name, with its suggestion of a time before white exploration or an uncharted place, ignored this long cultural history, as well as the ongoing struggle to define the urban landscape. When Seattleites created a new idea of what their city could be in the 1960s, their efforts to remake public green space expressed conflicting needs to assert order. Competing groups, including middle-class women activists, bird-watchers, Indians, and elites who hoped to educate and to shape urban citizenship, told stories that expressed fears and hopes for their collective and individual identities. By the 1980s, the struggle to shape a version of urban sustainability in Seattle revealed these tangled desires for the postindustrial city and the beginnings of an unfinished project of urban sustainability.

4

Ecotopia

In the spring of 1974 Richard M. Nixon stood on a platform on the glim-
mering Spokane River where he addressed a crowd of over 50,000 people at the opening ceremonies of Expo '74. Thousands of miles from the
scandal that dogged him in the other Washington, Nixon took refuge at
this mini–world's fair deep in the Republican territory, Washington's vast
agricultural hinterland once proudly dubbed the "Inland Empire." In his
speech Nixon told fairgoers, "The time has come to get Watergate behind
us and get on with the business of America." Spokane's local boosters
surely agreed with his sentiment. They themselves had put up at least $6
million to build the fairgrounds and bring in acts including the Carpen-
ters and Bob Hope. "Celebrating Tomorrow's Fresh New Environment" was
the vaguely hopeful environmental theme that boosters chose for the fair,
which they hoped would revitalize downtown and their fortunes. Expo '74
brought quickly dwindling federal urban renewal money to the downtown
business district, near the river, helping to clean up a polluted industrial
area for a waterfront park, a site that had suggested technology's failures.
In the 1970s Spokane was flush with a temporary boom in agriculture,
thanks in large part to Nixon's policy of selling wheat and other agricultural commodities to the Soviet Union. By the fair's opening day, however,
Nixon's farm policies had in fact already begun a decade-long process
that would undermine small farmers in favor of larger, consolidated op-

erations. Farmers in Washington, as in other parts of the country, had also begun to feel the brunt of the oil crisis. Eastern Washington, no less than Seattle, with its Boeing bust economy, felt the hard times.[1]

Spokane's modest fair contrasted sharply to Seattle's 1962 fair. The tumultuous years between the two world's fairs help to explain the starkly different utopian visions of urbanism in Seattle and Spokane. At the beginning of the 1960s Seattle's fair had exuded a kitschy, space-aged technological enthusiasm. Its high-modernist symbols still define the Seattle skyline: the Space Needle, the popular monorail, and a lavishly funded "science pavilion." Seattle's fair, the Central Association's miniature version of Seattle's possible future, celebrated the promise of Kennedy's "New Frontier": the corporate and government alliance of the military-industrial complex that made the Boeing boom possible. Even Elvis Presley came to town to use the exposition as the background for his film *It Happened at the World's Fair* (1963).[2] Demonstrating less enthusiasm, Expo '74's hokey environmental theme underlined the dampened trust in technology and, though only implied, a flaccid critique of the technological society that Seattle's fair never questioned. Even the fair's motto seemed to acknowledge that the damage had been done. Expo '74's lowered expectations paralleled the erosion of public faith in large-scale federal solutions, diminished urban spending, and a public's growing criticism of the enthusiasm for science and technology featured during the Seattle fair a decade before.

The fair embodied Nixonian environmentalism. By 1974 the environmental movement had made important strides, in no small part because of Nixon's efforts to co-opt a popular movement and outflank his Democratic political rivals. The president had signed into law far-reaching environmental legislation.[3] Nevertheless, most mainstream environmentalists viewed the fair's theme and the president's rhetoric with suspicion. They criticized the Expo for not living up to its muddled environmental image. Even the *New York Times* noted this co-opted environmentalism: "Visitors find displays that range from sobering messages of man's mutilation of nature to sales pitches for recreational vehicles." According to the *Times*, "no major environmental protection groups" showed up for the fair, and some called it a "disgraceful commercial sellout."[4] Fair officials felt that many of the environmental themes "distressed some visitors," and they placed musicians and clowns near "doomsday" displays to soften their effects.[5] At the same time, mainstream environmentalists complained that the fair's principal exhibitors—Ford Motor Company and General Motors—sent the wrong message for an environmentally themed fair in the middle of an energy crisis. With federal money no longer nearly as available as it had been in the early 1960s, the decentralized, local fair generated financing—as had early fairs—from Ford, Boeing, and other large corporations. In this business-dominated atmosphere, many environmental groups could not even afford booth space.[6]

Nixon, too, acknowledged the limits of the age. In particular, his speech touted a new urban policy, his disingenuous support of local initiative and smaller government: "New Federalism." According to the *New York Times*, the president had "recently been emphasizing governmental decentralization" in his speeches. The fair, an alliance of federal money and local power, had "not come from Washington," D.C., he said; "it came from Washington State."[7] That same year Nixon canceled the Model Cities program, which had done much to stimulate local initiative and decentralized decision making in Seattle's poorest neighborhoods. Despite his populist message of localism, Nixon understood local power as the power of local business interests rather than that of grassroots activists in environmentally compromised urban neighborhoods.

The Spokane fair reflected this historical context. To some it seemed like one more hollow public spectacle at a time when more and more people had begun to feel, for good reason, that they were on their own. Despite capturing the country's mood, then, Expo '74 may seem to be a mere footnote in the history of a fading world's fair ritual and certainly of little consequence to Seattleites on the other side of the state.

Yet this judgment is premature. While General Motors showed off its latest cars and fair managers smoothed over the sharp edges of their "green-washed" message, a group of counterculture environmentalists— self-identifying as bioregional utopians—quietly organized "a summer environmental symposium" at a nearby college as part of the Expo.[8] The symposium introduced important ideas and new energy to a movement that would soon wend its way toward Seattle, ultimately making a more lasting impression on Seattle's landscape and culture than any fair could. These ecotopians offered an articulate version of an emerging sustainability ethos that would remake the city and region during the 1970s and 1980s.

The low-profile summer series brought a variety of unconventional speakers to Spokane to discuss the state of the environment.[9] Unlike the mainstream environmental organizations at the fair, the panels assembled at the symposium represented what the historian Andrew Kirk terms "environmental heresy." By veering away from pollution, clean water, and other familiar environmental themes, these heretics offered an implied critique of mainstream environmentalists.[10] They were not necessarily hopeful. Indeed, they could be downright apocalyptic. Despite regulatory triumphs of the period, many environmentalists no longer felt assured of the planet's survival or of their own empowerment in the process. But their commitment to action and survival set the Spokane group apart. They stressed an everyday environmental politics, a do-it-yourself ethos that paradoxically echoed Nixon's message of small-scale, local, and low-budget solutions to seemingly intractable problems.[11]

Food constituted a principal concern. The true costs of its production and consumption in modern American society formed the center of their work for social and environmental change. A symposium entitled "Agriculture for a Small Planet" generated tremendous enthusiasm that summer, drawing more than mere hangers-on. A motley group of students, activists, and professors from nearby colleges, such as Washington State University and Central Washington University, as well as interested farmers from around the region, gathered to earnestly address a set of environmental concerns less obvious than open space, air quality, or animal habitat. Rather than advocate better government regulation or push for preservation of public lands, the group of counterculture environmentalists listened intently to, among other speakers, Wendell Berry, who wowed those gathered with his assertions that food is a cultural product, not a technological one, and that "urban and rural problems have largely caused each other."[12]

This last idea was particularly intriguing to activists at the fair who began to imagine an ambitious and holistic bioregional network linking hinterland places such as the Inland Empire to cities. Nevertheless, their approach to environmental and social problems may not have been quite as new or revolutionary as they thought. After all, the symposium's holistic ideas and decentralized activities shared important elements with the kind of thinking and organizing that people were already practicing in, for example, Seattle's Model Cities neighborhoods. Some of the activists who would help shape this new bioregionalism had also cut their teeth as urban community organizers earlier in the 1960s. Furthermore, these heretics reflected the shifting consumer ethos most clearly displayed in the antisuper-market rhetoric during the Pike Place Market protests, just as they shared with many Seattleites a growing desire to mend what they perceived as a broken bond between the city and nearby farming landscapes. Still, from unlikely origins at Spokane's fair, more than 280 miles from Seattle, a new idea of an ecotopian ethic began to form. Before it found a home in Seattle, this group would begin to articulate Seattle's emerging green ethos. In an era of limits, these activists hoped that their new politics of production and consumption might bridge the divide between producer and consumer, city and country, culture and food.[13] And like Mark Tobey, with his vision of the Pike Place Market, they hoped to reestablish a more just and democratic city around the marketplace and the production of food. At mid-decade they did this using the guise of ecology and the earnest but open style of the counterculture. Not long after Expo '74, they took their brand of ecological activism to Seattle, making the city the center of an ecotopia.

Inventing Ecotopia

Mark Musick, a Washington native and Central Washington University graduate, was a key organizer of the summer institute that brought dozens of speakers to Spokane. Musick's interest in organizing and grassroots politics grew from his previous summer's experience as a VISTA volunteer and community organizer in Denver. There, Musick trained in the Saul Alinsky style of community organizing. During his time in Denver he had an experience common among many white, middle-class civil rights organizers in the late 1960s. Musick recalls that an important Chicano activist told him and other organizers, "We don't want you in our community; we want you to go home and get your community off our backs." For Musick the conversation was a turning point. "We can't change the world," he said, "but we can change ourselves and our little corner of it." After Denver, he returned to Washington State armed with organizing principles and prepared to apply them to a new project.[14]

Musick and his friends and future comrades Woody and Becky Deryckx were inspired by the speakers at the symposiums, especially Wendell Berry but also Peter Caddy, from the famous Findhorn ecovillage in Scotland, and Anthony Judge, of the International Union of Associations. A pioneer in computer database and network development, Judge inspired the group with his presentation about "facilitative processes" and "network" language that could be used in organizing dispersed groups and resources. Musick remembered that these "network strategies for social change" were crucial in imagining how to bring "together like-minded people to work for what you . . . want" rather than just fight against the system. In intense conversations throughout the conference, Musick and others began to consider how they could use what they learned that summer to create their own alternative networks in the region.[15] All they needed was a push.

Wendell Berry provided the push. The famous writer began corresponding with the activists who had coaxed him from his Kentucky farm to speak at the Spokane symposium. On July 4 (Musick emphasizes the significance of the date) he wrote the group a letter in which he suggested that their experience and contacts at the fair gave them "a peculiar usefulness to the environmental movement." Berry asked the would-be organizers about holding another symposium that could "bring together the various branches of agricultural dissidence and heresy." Such a symposium, Berry thought, might include farmworkers, urban consumers, organic gardeners, conservation groups, and the publishers of works on dissident agriculture.[16] Berry especially encouraged them to address what he said was an "urgent need" for a forum in which urban environmentalists could "begin to learn something about farming."[17] Musick remembers that Berry's letter "was like a match thrown on dry tinder," sparking "a flurry of organizing throughout the region."[18]

Musick and his friends would later find a name for their organization in the word *tilth*, meaning soil prepared for cultivation. The word evokes agriculture—the soil necessary for laying down good roots—but also suggests a sense of renewal and the idea of the rooted regional connections they hoped to make. Tilth, like many other dispersed counterculture environmental projects of the period, was flexible and pragmatic. Far from a group of back-to-the-land dropouts, these would-be ecotopians envisioned a postindustrial "whole earth ecology" vision for the region, one that explicitly incorporated the city and consumers as part of an alternative network of production and consumption. Food was their priority, but food illuminated a new kind of environmentalism.

To understand the adaptability and pragmatic nature of this movement requires an explanation for the intellectual foundation, practical models, and bioregional myths that underlay Tilth. Although the collective experiments that eventually would make up Tilth's counterculture landscape

differed widely in specific emphases—organic agriculture, co-op organizing, alternative energy, or urban land reform—they shared some common
assumptions that set the counterculture environmental movement apart
from much of mainstream activism.[19] Tilth saw self-sufficiency as the root
of survival. Food and energy were intertwined components of this vision.
Bringing a sense of coherence to these dispersed projects was an often-
emphasized and almost mysterious-sounding goal of achieving "steady
state" ecology in the Northwest. The vaguely mystical yet scientific phrase
shared precedents with the rhetoric of other reformist and utopian groups
in American history that sought restored balance with nature. In particular, ecotopians were inspired by the blend of practical science and political
mission they found in the work of the ecologist Howard T. Odum.

Odum was a leading voice in the new science of "human ecology" during the 1970s. Brother of the more famous ecologist Eugene, Howard published his early ideas about the "energetics of ecological systems" in *Environment, Power, and Society* (1971) and later in the more activist-oriented
book *Energy Basis for Man and Nature* (1976), cowritten with his wife, Elizabeth Odum.[20] In his introduction to *Environment, Power, and Society* Odum
described the contemporary human ecology movement as the "widespread
efforts under way in undergraduate colleges to develop courses in human
ecology that are pertinent to a new generation grappling for survival on the
planet."[21] Urging the study of "limited energy of environmental systems,"
Odum encouraged activists to link the more abstract idea of environment to
the concrete and practical world of social change. In this way, counterculture environmentalism would have the potential to overlap with a variety
of social movements when the organization came to Seattle. His notion of
"system" encompassed "small ecological systems," "large panoramas that
include civilized man," and "the whole biosphere."[22] At the core of Odum's
thought was the idea of creating a more realistic energy economy based on
an honest accounting of systems and their operations. A critic of economies
based on fossil fuels, Odum argued that all "critical issues in public and
political affairs of human society" had this "energetic basis," an idea that
naturally appealed to ecotopians during the mid-1970s oil crisis and em-
bargo.[23] Odum's human ecology meant making interventions at various
scales of politics, consumption, and energy production that could bring
balance to the various subsystems and habitats within a region.

Urban activists found inspiration in Odum's ideas, especially the macro and micro views of the system that engaged "the big pattern" of the ecosystem and the constituent parts that defined the system, including "the forests, the seas and the great cities."[24] Perhaps the metaphor most powerful for understanding Odum's ideas about ecology was the example of the terrarium, which he often used to illustrate the idea of a "closed system" in which one could trace the ways energy and materials were "cycled and reused."[25] Unlike the large and fantastical landscapes produced on the grounds of the world's fair, Odum's energetic systems ecology, his scientific explanations for societal problems, and the persuasive comparison of systems to terrariums and aquariums provided a way of thinking about environmental problem solving on a more immediate level—neighborhood by neighborhood. In this formulation, urban, rural, and wilderness spaces could be understood as part of a constant exchange of information and energy. Mark Musick's and Tilth's mission, then, was to regain a balance within this system in any ways that seemed appropriate.

The most famous of these experiments and an inspiration to Tilth activists was the West Coast branch of the New Alchemy Institute, a group funded by Stewart Brand's Point Foundation.[26] New Alchemy provided an example of human ecology in practice. World's fairs had in the past often been testing grounds for urban planning and utopian cities; the New Alchemists served as much the same, though in a decentralized, small-scale version.[27] They enacted and tested Odum's ideas, modeling an experimental ethos and ethic of self-sufficiency that Tilth and ecotopians would emulate as they invented their own bioregional ethic. In "A Modest Proposal," the group's founding manifesto, its leader, John Todd, described New Alchemy's vision of a biotechnology that could stop the alarming loss of both human and biological diversity. Todd described a mission that included creating systems for survival—such as the group's famous "backyard fish farms"—that were simple and could be used by the poor. The group advocated development of local "eco-economies" and the "evolution of small decentralized communities, which in turn might act as beacons for a wiser future."[28] Such messages, delivered in book titles such as *What Do We Use for Lifeboats When the Ship Goes Down?* resonated with the counterculture's simultaneously apocalyptic and creative mood. In a time when many saw impending ecological disaster, the New Alchemy Institute inspired do-it-yourself terrarium building on a dispersed and unprecedented scale throughout the country.

By the mid-1970s the counterculture environmental movement was emphasizing the development of an increasingly regional and locally appropriate set of practices. Such ideas were advertised and circulated in the burgeoning underground press of the period. In 1974, for example, the Portola Institute, in Menlo, California—the same organization that helped make the *Whole Earth Catalog* (*WEC*) possible—published its *Energy Primer*, a book similar to the *WEC* but focused exclusively on technologies "geared to the local conditions of climate, economy, geography, and resources."[29] Along with running detailed articles describing how to construct windmills and aquaculture ponds, the *Primer* listed in its back pages various similar efforts and communities up and down the West Coast. Such organizations invariably expressed a desire for "communicating and sharing information."[30] The proliferation of regional *WEC* knockoffs was an indication of this flourishing movement. Some were fly-by-night and some were more enduring, helping to knit together regional networks of alternative scientists and technologists and especially a nascent bioregional food movement.[31] From Odum and New Alchemy, ecotopians would learn how to raise algae in backyard vats to feed tilapia, how to raise organic vegetables with composted wastes, and what kind of energy and effort these projects required. The ecotopians of the Northwest were unique, however, for their ability to organize these dispersed islands of activity into a regional alternative system of production and consumption accompanied by a useful mythology of place.

Unlike many other dispersed counterculture groups, by the mid-1970s Tilth had a powerful and useful origin myth that paralleled its work. The movement's bioregional focus was encapsulated in the Ernest Callenbach book *Ecotopia*, part fantasy novel, part chronicle, and part practical guidebook. Mark Musick recalled that *Ecotopia* simply affirmed the work he was already doing in the Northwest. The fantasy of escape in difficult times resonated with many. *Ecotopia* imagined a postindustrial era in the near future when daylighted streams would again flow through downtown streets and vacant lots would be reclaimed for agriculture. *Ecotopia* introduced the fantastical idea of the Northwest, having seceded from the Union, transforming itself along ecological lines. Callenbach's vision offered escape for some. (But perhaps not for others: in an era of racial tumult, Callenbach pictured his Ecotopia as a racially divided postapocalyptic society, suggesting a darker side to the new green movement.) Though secession may have been fantasy, many refugees from the Bay Area and beyond had begun moving to northwestern cities to find or create this lifestyle on the cheap. In significant ways, Callenbach's fantasy paralleled the journey of many activists who settled in Seattle and other northwestern cities in the 1970s.[32]

To complement this mythology, activists on the West Coast—including Callenbach himself—created their own *WEC*-like sourcebook called *Seriatim: Journal of Ecotopia*, "a journal about the Ecotopian bio-region," which elaborated on the book's core ideas. In the 1977 inaugural edition the editors wrote: "The novel Ecotopia[,] by our friend Ernest Callenbach, was influential in getting us started. We liked a number of his ideas and liked his optimistic approach to the future of the Northwest."[33] Like Callenbach and other ecologically minded activists, the editors of *Seriatim* shared a "vision of a self-reliant, small-scale, stable state series of systems, through all the facets of our lives." *Seriatim's* editors hoped to "document and foster the growth" of an "environmentally-attuned stable-state society" that they claimed was emerging "in the Northwest corner of the continent." Each issue contained articles in departments that included agriculture, alternative technology, "shelter," forestry, and "gatherings." The articles ranged widely, from descriptions of low-energy techniques for timber harvesting to discussion of household ecological consciousness and the development of a recycling network in Portland, Seattle, and elsewhere in the region. The journal, according to editors, would "focus to a large extent on reports of research and innovations of relevance to the Ecotopian enterprise."[34] Unlike other resource catalogs and how-to manuals of the counterculture environmental movement, *Seriatim* helped to perpetuate a distinct sense of regionalism, the explicit idea of steady state ecology, and a spirit of experimentalism in a variety of locales that made up the "Ecotopian bio-region."

With Odum's intellectual foundation, ample examples from the West Coast counterculture movement, and a sense of their own mythology of place, Musick and the Deryckxs decided, in true 1970s fashion, to organize yet another conference modeled after their positive experience at the Spokane Expo earlier that summer. The "Northwest Conference on Alternative Agriculture" would "assemble the ingredients for a debate" and "promote a higher level of communication, understanding, and cooperation among people . . . exploring innovative alternatives in agriculture in the Northwest," announced the organizers. The preparations for the conference re-

SERIATIM
Journal of Ecotopia

Autumn 1977 $2.50

CREDITS
Cover:
Illustration by Julie Reynolds
Back Cover: Raphael

Seriatim is a journal about the Ecotopian bio-region located in the Northwest corner of the American continent where an environmentally-attuned, stable-state society is emerging. Seriatim aims to document and foster the growth of that society.

We focus to a large extent on reports of research and innovations of relevance to the Ecotopian enterprise. Our definition of research includes software systems — trade networks, economic alternatives, strategies for personal and social change — as well as hardware — renewable energy applications, agri/aquaculture methods, recycling technologies.

Seriatim — (sury-ah'-tum) — is an obscure adverb meaning "in succession; one thing following another in natural order, as in the growth of a crystal structure or a succession forest."

Managing Editor: Bruce Brody
Co-Editors: Frank De Santis
         Don Bright
         Patti Clemens
Typography: Type & Graphics,
         San Francisco
Printing: Fricke-Parks
Special Thanks: Shoshona
         Mike Di Palma
         Richard Feldman

SERIATIM
Journal of Ecotopia

Volume 1, Number 4   Autumn 1977

Seriatim: Journal of Ecotopia, ISSN 0147-2275, Issue No. 4, Autumn 1977. Published quarterly by Seriatim Magazine. Office of publication: 122 Carmel, El Cerrito, CA. 94530. Subscriptions: $9/yr. in Ecotopia, $12/yr. elsewhere, $12/yr. institutions. Postmaster: Forwarding and return postage guaranteed. 2nd class postage paid at El Cerrito, CA. 94530. Copyright ©1977 by Seriatim Magazine. All rights reserved.

Table of contents of *Seriatim: Journal of Ecotopia* 1, no. 4 (Autumn 1977), with examples of the bioregional viewpoint expressed in the journal. Tilth Association Records, Cage 602. Courtesy Manuscripts, Archives, and Special Collections, Washington State University, Washington State University Libraries, Pullman, WA.

141

flected a focus on alternative agriculture, one born from the wheat fields and agricultural universities around Spokane but that depended on urban consumers and activists.[35]

Conference organizers hit the road in late summer of 1974 to visit some of the far-flung corners of this emerging ecotopian landscape. From Ellensburg they traveled over the mountains to Seattle to meet with the "Cooperating Community," a Seattle co-op consortium that included representatives from "all the co-ops and collectives in Seattle." They then made the four-hour trip to Portland, where they met with Lee Johnson (one of the founding members of RAIN, a group dedicated to appropriate technology in the Pacific Northwest) to discuss making contacts in Oregon. Later that day the entourage visited Portland's "Apocalyption Reconstruction Company," a group teaching "homesteading skills for urban YMCA kids." The leader of the group agreed to give a talk at their conference. The preconference road trip indicated the geographical reach of the emerging countercultural movement and the different interests involved: from marketing and consumption, to self-reliance and alternative technology, to urban homesteading and farming. The conference quickly broadened to include a variety of environmental heretics, not only farmers. The territory involved extended from Canada to California, but given that the group defined itself in bioregional terms, it was most concerned to notify individuals and organizations in Washington, Oregon, Idaho, and British Columbia, with Seattle and Portland emerging as key nodes in this hinterland/urban circuit. When some in the emerging network expressed the idea of renaming the upcoming conference the "Northwest Conference on Rural Alternatives," Deryckx argued that the theme need not be limited to "rural activity," a point that he said was "brought out strongly" whenever he visited "an organic community garden project on a formerly vacant lot in a city or town."[36] Organizers encouraged "conservationists of both city and country" to attend, as well as people interested in alternative energy and food distribution systems. Many of the enterprises were bound to fail. But to make any of them work, nobody could be left out of this network, which required a symbiotic bond between a growing market of enlightened urban consumers and organic gardeners as well as rural producers. Entities such as the Pike Place Market thus became ever more important, symbolically and concretely, for healing the rift between city and hinterland during the 1970s.[37]

According to Musick, "people came out of the hills" to attend the Northwest Conference on Alternative Agriculture.[38] Holding the event during November 1974, the main organizers first began officially using the name Tilth for their loose and evolving organization. Using word of mouth, letters, and little else, organizers were able to draw nearly eight hundred cooperative members, "tool freaks," and farmers to the event, which they billed as "a forum for people seeking alternatives to current agricultural practices and attitudes."[39] According to the organizers, the conference brought together "outstanding leaders from throughout the nation" to help "clarify the issues and chart the possibilities for the integration of agriculture and the environment, for land reform, and for economic stability and justice."[40] Far from a loose counterculture "happening," then, the conference exhibited a high level of organization and intention, gathering together the threads of the region's counterculture economy, a diverse and mainly urban-based constituency. For example, Tom Allen, from Bellevue, a suburb immediately adjacent to Seattle, brought a curiosity about the "development or discovery of existing plant genotypes suitable to cold inhospitable environments"; the Earth Cyclers, a group in Edwall, Washington, brought their interest in "methane generators, trading networks, organic farming, wind generators, solar collectors, electric motors, oxen, [and] videotape." Gene Kahn, of Cascadian Farm, came to talk about organic vegetables, specialty crops, and small grains. Seattleites Dan Carlson and Shari Reder wanted a "way to maintain an urban-rural balance and healthy independence," while Bill McCord hoped for "urban-rural connections" that could be "lasting and directed towards long-term reform." Community Produce, of Seattle, was "interested in *produce marketing*," and members of Seattle's Capitol Hill Food Co-op said they had come to distribute "natural food from small farmers and independent producers." And just about everyone stressed "communities" and "connecting."[41]

One of the "outstanding leaders" invited to speak at the conference, which was held at Ellensburg, was Richard Merrill, of the New Alchemy Institute, the influential group that sought a world of "decentralized technology based on technological principles," suggesting the conference's mixture of interests in food culture and appropriate technology.[42] The author of *Radical Agriculture*, Merrill specialized in alternative agriculture as an aspect of survival. His keynote address stirred the crowd of "mavericks and dissidents" with his idea that "land, people, energy, and food" are "the basis of life."⁴³ Merrill spoke of the necessity of maintaining quality fertile soil in the face of fossil fuel–based economies of scale that used "chemicals, machinery, monocultures, and hybrid crop strains." Such evils, he argued, ultimately threatened soil, wildlife, public health, and rural economies.⁴⁴ An experienced activist living in California, Merrill keenly understood the alternative food economy and the obstacles that Tilth faced. The "final barrier to alternative agriculture," he concluded, was the "isolation . . . between urban and rural cultures." Merrill might well have pointed to Seattle and the Pike Place Market when he said, "Today most problems are 'urban' problems and we seem to miss the point that the decaying urban condition has its origins in the decaying rural condition. A radical agriculture, geared to the needs of a post-industrial society, must begin, in the cities as well as the farms."⁴⁵

The conference events reflected this urban focus as well. One workshop included presentations on renewable sources of energy from Ken Smith, of the "Ecotope Group" in Seattle, and Lee Johnson, of the Energy Center at the Oregon Museum of Science and Industry in Portland. Friday featured a discussion of "ecological building practices" given by Len Dawson, from Seattle's "Housing Assistance Group," and talks on water, wind, and solar power. On Saturday the conference concluded with panels entitled "Land Reform, Land Use Planning, and Rural Communities" and "Alternative Marketing, Economics, and Finance." A "transformed social order," these activists argued, would require "radical solutions," including "new patterns of land ownership and marketing."⁴⁶

Economy and ecology were closely linked in this understanding of environmental change—a counterculture and a counterconsumer economy. For instance, Merrill stressed the role of "second wave" food co-ops and farmers' markets in making the coming revolution possible and in shifting the "grassroots food economics from inner-city radicals to broad-based support groups." Ecotopians were willing to engage with capitalism and markets in this transformation; they simply hoped for a more manageable and local economy, one that more accurately and honestly accounted for the costs and benefits to the overall environment, including the cost for human communities. Merrill beseeched his audience to "create a local, and endemic, regional agriculture," an "urban agriculture" of integrated farms and co-ops.⁴⁷ By 1974 it was already clear from Musick's preconference tour that people working at various scales in the Northwest were busily creating these connections, and most of them were based in cities. In its capacity as an umbrella group, Tilth would make these connections manifest. At the same time, by moving the focus of organizing toward cities and showing the urban connection to hinterlands, Tilth would join activists in Seattle already working for stronger and more sustainable urban communities. Organizers and participants at the conference found their central theme in food and energy, but they imagined a greater Northwest of communities reestablished "in harmony with the earth." And they advocated these new approaches "in context with the systems of energy and the environment."[48] As Merrill emphasized, returning to the land was a process that could happen anywhere, especially in cities.[49]

Ecotopia in Seattle

As the Ellensburg conference suggested, the largest constituency of ecotopians in the Pacific Northwest resided in cities, particularly Seattle. Tilth found that the core of ecotopia consisted of committed activists ranging from community organizers to co-op members. One of the first events announcing Tilth's arrival in the city was held in 1976 on, appropriately enough, the grounds of the Century 21 fair. At the time of this teach-in, called "Tools for Transition," Tilth was still a regional umbrella organization with no permanent home in the city. The seminar met in the building that had housed the American science exhibits at the 1962 fair, where Seattleites had celebrated the new frontier, space travel, and a high-modernist utopian urbanism. But the Tools for Transition seminar addressed a different future for the city. Over several weeks the activists and the public met to discuss the "alternatives being created for a solar energy-based, steady state society." Such a society, organizers argued, would necessitate "new approaches in energy, agriculture, economics, health, communications, education, and community building," together forming the foundation of "a more positive future."[50]

Tools for Transition inhabited the shell of the 1962 world's fair among the space-age relics, a reminder of earlier efforts to order and revivify the shifting metropolis in the age of early 1960s urban renewal.[51] The Seattle fair had evolved, in part, from the concerns of the downtown Seattle businesspeople who had hoped to develop the Pike Place Market in the late 1950s and early 1960s and to maintain downtown's central economic position in the rapidly shifting region, their modernist hopes leading them

Founding meeting of Tilth at Pragtree Farm, 1977. Tilth Association Records, Cage 602. Courtesy Manuscripts, Archives, and Special Collections, Washington State University, Washington State University Libraries, Pullman, WA.

to hatch their plan for a fair, a theme park–like compound on the northern edge of the city and an enduring cultural center of the state.[52] Although the fairground succeeded in defining the northern edge of downtown and the skyline of Seattle, the effort did little to stop the centrifugal pull of subsidized suburbs. The year of the fair saw Seattle's population near an all-time peak at 557,087, but by the 1970s and Tilth's arrival, the city had lost an estimated 25,000 residents, many of them to the former farmlands of immediately adjacent suburban King County, which grew by over 215,000 during the decade.[53] By the middle of that decade, the scope and power of postwar urban renewal and redevelopment schemes had diminished substantially.

Tilth and the network of counterculture green consumers helped to fill the vacuum left by vanishing 1960s urban renewal programs, offering an appealing low-budget alternative in stingier times. Unlike the broad strokes of urban renewal and the utopian images of world's fairs past, Tilth's decentralized, small-scale utopian landscapes would play out on a neighborhood scale at the neglected edges of downtown, filling the gap

146

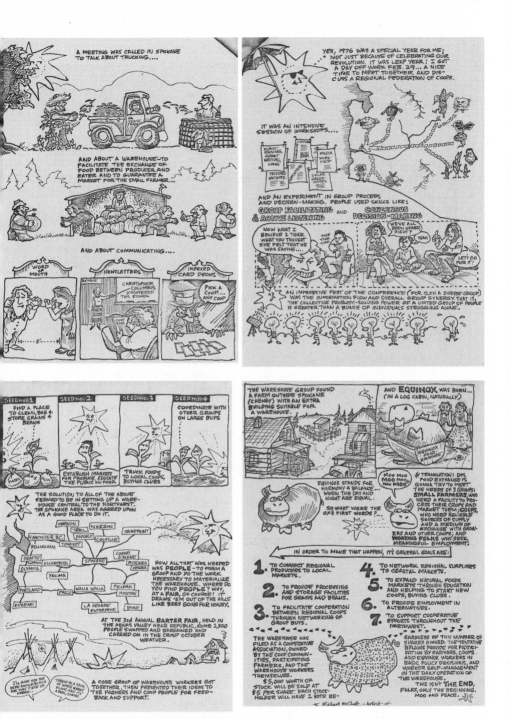

Cartoons depicting the invention of a bioregional community and the proposed development of regional markets of growers and consumers in the larger Ecotopian region, including a map of the network. Published in *Tilth Newsletter*. Tilth Association Records, Cage 602. Courtesy Manuscripts, Archives, and Special Collections, Washington State University, Washington State University Libraries, Pullman, WA.

Barter Fair and gathering, 1977. Tilth Association Records, Cage 602. Courtesy Manuscripts, Archives, and Special Collections, Washington State University, Washington State University Libraries, Pullman, WA.

left by years of downtown-focused planning. Urban pioneers and opportunists, Tilth's ecotopians responded to the city's demographic and social landscape with a reform agenda tying the urban to the rural. In the cities' deterioration they saw an indication of a more profound imbalance, but armed with Odum's ideas, they addressed this imbalance with a faith in human ecology and hoped to make their politics tangible in the built environment of Seattle neighborhoods. At the same time, however, the previous decade's social movements and community activism—especially the ground broken in the market conflict and in the Model Cities program, as well as the shifting attitudes of local officials—framed what was possible. Tilth's emphasis on community agriculture in cities, perhaps better than any other aspect of its activism, allowed this counterculture green movement to find a foothold in the strong neighborhood-based political activism in post-1960s Seattle.

   In the process, the garden in the city became the enduring symbol linking urban rebirth, food, and nature during this period. Amid the economic stagnation and reduced spending on cities, ecotopians, like their fictional doppelgängers in Callenbach's novel, found fertile ground in the marginal inner city and in the neglected first-ring suburbs of Seattle.

In Seattle, three landscapes and the people who helped build them best display this influence. Ecotopians began building one of the nation's largest community gardens, the P-patch community garden, in one of only a few such programs to be managed by city government and therefore institutionalized as part of an urban bureaucracy. During the 1970s, Tilth also created a public educational center—along the lines of New Alchemy—headquartered in the Wallingford neighborhood on the contested grounds of a historic school. Finally, in an example of the crucial link between countercultural activism and the community empowerment movement in the Central District, Seattle's ecotopians helped to build the Danny Woo Community Garden on an urban hillside in the International District.

Each story concerns an enduring landscape that has left a permanent mark on Seattle's urban politics and sense of community; each site constitutes an example of small-scale, low-budget, grassroots urban renewal in the 1970s; and each landscape is related to food and urban agriculture. Akin to utopian fairgrounds, these more humble sites served a didactic purpose as well, educating the public about food and culture through the spectacle of production and especially new forms of local consumption. Each of the three stories also suggests the influence, diversity, and lasting effect that ecotopian activists would have on the city's tastes, style of politics, and sense of community.

The P-patch

The story of urban agriculture in Seattle did not begin with Tilth and ecotopians, yet the ecotopians who would join Tilth's regional network contributed expertise, financial support, and the marketing necessary to build a gardening infrastructure in the city. With its mild climate and extended growing season, Seattle has long been a city of backyard gardeners. At the beginning of the twentieth century Seattle had joined the rest of the country in its embrace of the City Beautiful movement and Progressive Era parks. The Olmsted brothers had helped Seattle with the design of greenways and even public gardens, yet, like those in other prewar cities, its productive food gardens existed mostly in small private plots primarily on the immediate outskirts of the urban core in first-ring suburbs. Before the 1950s the city had obtained most of its fresh commercial produce from the extensive nearby truck farms to the far north of the city and in the fertile valleys close by.[54] This evident decline of small local farming and the consolidation of the farming industry motivated early Pike Place Market preservationists and their grassroots consumer movement, and the same impulse continued with ecotopians.

The history of gardening as a visible tool of urban reform and education in Seattle is less well defined, however. In the early twentieth century, the Seattle school system used agriculture to teach children about nature and inculcate values of citizenship.[55] As early as 1971, in the guise of reform, the municipal government and neighborhood activists employed agriculture as part of the Yesler Atlantic Neighborhood Improvement Project (YANIP), a War on Poverty community organizing effort meant to rejuvenate the predominantly African American neighborhood near the Central District. In operation during the Model Cities period, YANIP initiatives encouraged a variety of neighborhood projects, including street improvements, a childcare center, and a neighborhood youth corps summer employment program. In 1971 the organization sponsored "Environmental and Secretarial Aide Programs," which offered "an outdoor work experience for the males, and office practice experience for the females." The young men in the program, according to the *YANIP Newsletter*, "contributed their youth and strengths to the cleanliness of the Yesler Atlantic Project" area in 1971 and "planted, cultivated, and harvested" a vegetable garden on a vacant lot between some older residential homes in the neighborhood.[56]

Photographs accompanying reports on the efforts show young men leaning on their garden implements among orderly rows of lettuce, tomatoes, and cabbage.[57] When people in the neighborhood vandalized the gardens, the city dropped the project. But other attempts at gardening would follow, and in the early 1970s YANIP encouraged private gardens and yard maintenance, providing gardening tools and know-how to residents. The neighborhood also hosted an "inner city farmers market" sponsored by a group called Hunger Action Network. These early examples of agriculture-as-reform and as an antipoverty measure set a precedent for a decade of furious urban agriculture expansion to come. Tilth followed in these footsteps.

While YANIP activists taught young men about gardening in inner-city plots, on the other side of the city, just north of the University of Washington, a young woman named Darlyn Rundberg began building a unique urban farming experiment in the residential neighborhood of Ravenna.[58] In 1970 Rundberg was a student at the University of Washington and lived just north of the University District in the neighborhood that was also home to the Puget Consumers Cooperative, the longtime Seattle cooperative grocery store and a hub of counterculture environmentalism and alternative consumption. A co-op member, Rundberg lived near an agricultural anomaly in the middle of a residential neighborhood: a three-acre former truck farm, owned by Rainie Picardo, which was all that remained of a twenty-acre family farm, one of many that had existed on the other side of the old city line before the 1950s. Like the Kent Valley, the north end of Seattle near the University District had then contained large truck farms. The large shopping center called University Village sits on top of one of these once prosperous Japanese truck farms. By the late 1960s Picardo had decided to retire, and the old farmer allowed Rundberg and other neighbors to garden portions of the rich farmland in small plots. In the early 1970s, however, the pastoral scene was in jeopardy. Following a common pattern, the city raised Picardo's taxes, and the erstwhile farmer was forced to sell off parts of the land for a parking lot to meet his tax bill. It appeared that the rich soil Picardo had tended for most of a century would be lost. (The community garden program to come memorializes Picardo's name in the *P* of "P-patch.)[59]

For Rundberg and others, the imperiled farm was a clear example of the stakes in the environmental movement. In 1971 she contacted the city council to suggest that it preserve the land to "provide publicly owned allotment gardens within the City of Seattle and to preserve the P-patch's three acres of prime agricultural soil for that purpose."[60] Her request fell on receptive ears. A revolution in city politics during the period of the market preservation battle brought a younger and neighborhood-oriented council to power in the 1970s. Open to change, they considered Rundberg's idea, and a young self-described libertarian city council member named Bruce Chapman—who would later serve as Ronald Reagan's deputy assistant—was an important champion of the plans.[61] In "A Look at Seattle's Back-to-the-Earth Plan," the *Seattle Post-Intelligencer* described the proposal, joking, "The next change to Seattle's skyline may be ears of corn between the Seafirst Building and the financial center as Seattle possibly turns back to the earth."[62] Of course, such headlines delighted the city's ecotopians, who had just that in mind.

By the 1970s Seattle sustained a variety of established and newly invented cooperative enterprises that served as an infrastructure of support for all sorts of alternative undertakings. Setting the tone since the 1930s, Seattle's REI co-op sold outdoor gear to an ever-expanding base of outdoor enthusiasts. Membership in REI was almost a prerequisite for newcomers to the region who hoped to bond with their new home through the purchase of itchy wool socks and "waffle-stompers." Seattle also had one of the first HMOs, the Group Health Co-operative, which began in the 1940s to serve Seattle's health needs within a co-operative concern owned by both physicians and patients. But it was the North Seattle Puget Consumer's Cooperative (PCC)—"the food co-op," as it was called in the 1970s—that served as perhaps the most important magnet, drawing together a growing group of counterculture environmentalists and concerned consumers who shared Tilth's beliefs about the link between food and politics.

The PCC antedated the "new wave" of food cooperatives by many years.[63] The original co-op began in the 1950s as a food-buying club established by like-minded health food activists. Early on, the club held its business meetings at the University Friends meeting center—a Quaker hall near the University of Washington—but most of the co-op's activity was conducted in the living rooms of its two hundred members' homes as late as 1965. According to the co-op's newsletter, these consumers were "fostering healthful and sensible consumption and providing more healthful goods," as well as "introducing biodegradable detergents" that would not "pollute rivers, soil, etc."[64] In 1967, the co-op had delivery and pickup points in most of Seattle's neighborhoods—including both Model Cities neighborhoods and counterculture areas, such as Madrona, Montlake, Capitol Hill, Northlake near Fremont, and Ravenna in the north end. The co-op newsletter was the main link for consumers and announced prices and availability for wheat germ, carob, honey, and other bulk staples. It also recited practical business information for its members.[65] By the time the PCC opened its first storefront, in north Seattle's Ravenna District (within walking distance from Picardo's farm), it was one of a growing number of co-ops near the centers of youth culture (e.g., the neighborhoods around the University of Washington and Capitol Hill). These other, newer co-op businesses included Black Duck Motors, a co-operative bakery called the Little Bread Company, and a group that called itself Alternative Finance. Seattle's co-ops in the 1970s had their counterparts in a growing array of local and statewide suppliers, the same hinterland network represented at the Ellensburg conference, including Okanogan Honey and Methow Valley Food Co-op in the northeastern part of the state and the Snohomish Food Co-op north of Seattle.[66] The growing membership of the PCC supported an alternative food economy in Washington State, helping to link the choices about consumption in the urban neighborhoods with Seattle's hinterlands.[67]

With the north Seattle store up and running after 1970, the PCC's membership expanded quickly, and so did the scope of the co-op's environmental consumer politics. With each newsletter after 1970 the PCC reflected a merging of concerns about environmentalism and consumption. Although less businesslike than the earlier bulletin, and often humorous in tone, the post-1970s newsletters of the growing organization indicated the concerns of a mostly sober and thoughtful group. Long before anyone had uttered the question "paper or plastic?" PCC members had agonized about providing paper bags to customers (or making them bring their own), and they labored to find group consensus at interminable member meetings. But the dominant concerns revolved around ecology, consumption, organic produce, recycling, and the development and use of an alternative network of small local produce and product providers.

The co-op, then, was more than a store selling bulk staples to the community; it was a powerful political force supporting inventive projects and endeavors related to the politics of food. Rundberg was able to draw from this community for help when the Picardo farm appeared threatened. Eventually the city agreed to pay taxes on the farm for the next two years, and the city council allowed the Puget Consumers Co-op to manage the gardens as a pilot project supported by member dues. The city had made its first foray into the agriculture business. According to the agreement, the counterculture environmentalists at the PCC and P-patch neighbors could apply for plots. The PCC would help them get started. During spring planting at the P-patch, Koko Hammermeister, the PCC's project manager, announced to the *Seattle Post-Intelligencer* that "organic gardening [would] be practiced" on the north Seattle site and that the co-op would "assist those . . . new to the concept." The P-patch would be "free of synthetic materials" and was, according to Hammermeister, "in a very delicate organic balance" that, she said, "must not be ruined by abusive chemical technology."[68] The P-patch was the first of many creative counterculture experiments in liberating land from the market, building a smaller decentralized community context, and educating the public about organic produce and the ideas of ecology.

Co-op members increasingly involved the PCC in bringing an awareness of organic agriculture not only to consumers but also to a swelling community of backyard vegetable gardeners, offering them guidance and inspiration. The co-op funded a unique research project that explored techniques for growing winter vegetables in the Northwest and educated its members in the ways of composting. The project was a joint project of Tilth's Pragtree experimental farm—and especially the work of Binda Colebrook—and the cooperative, an example of the inventive and regionally appropriate work that was central to the group's ethos.[69] After 1970 the newsletter also began to provide garden tips and offered seeds at prices "considerably below the usual Technicolor seed speculators."[70] According to Hammermeister, the P-patch provided a "space for people to grow their own food and get in touch with the earth, experiences that are not usually possible in an urban environment."[71]

With a growing community of consumers in the mid-1970s, the PCC became a clearinghouse for ideas and a forum for action helping to give shape to Seattle's counterculture environmental movement—the environmental heresy that Wendell Berry had described. The store eventually helped bring once fringe ideas into the mainstream of urban consciousness. Throughout the 1970s, the PCC helped Seattleites learn to link consumption in their private lives and homes to a larger picture of a more humane and ecologically healthy urban and rural landscape—a landscape achieved through their everyday choices about what they ate, bought, grew, composted, and recycled.

By 1974 the city was satisfied with Rundberg's efforts and impressed with the obvious public enthusiasm for the project. That year the city council considered creating a new citywide P-patch program, named for Picardo's original acreage. Seattle was a "city of industry, shipping, and education," according to the Seattle Times, but "may have grown too far away from the good earth." The local newspapers detailed the efforts of two council sponsors and made fun of the novel idea that "Seattleites requaint [sic] themselves with the soil by growing gardens on vacant parcels of land owned by the city."[72] A few days later the city council officially authorized the Department of Human Resources and the parks department

The original P-patch in north Seattle, 2010. Photograph by Jeffrey Sanders.

to create the garden project "for recreation and open space purposes."[73] On March 14 the *Seattle Post-Intelligencer* reported that the proposal for the new city program had "sailed through" to approval. The city's coordinator for the new P-patch program remarked, "To my knowledge, this is the first time that a city has acted to preserve agricultural land in an urban area for agriculture. This could be a major precedent."[74] Soon after, the program expanded throughout the city, increasing from 500 to 1,000 plots, and thus began the citywide P-patch program and Seattle's evolving experiment with grassroots ecotopian urban renewal in the 1970s.

Whether they meant to or not, Rundberg and the PCC's membership successfully drew the city into the farming business, an early example of counterculture ideas finding unlikely support from a city bureaucracy. The PCC, activists, and city workers combined their energies to build momentum behind community gardens. With the bureaucratic machinery in gear, various city departments, city workers, and citizens were suddenly on the prowl, eyeing vacant plots and unused parking lots for their potential to be liberated for food production. City workers exchanged memos about suitable vacant spaces for farming.[75] One memo suggested how to "locate a land site," giving examples of land to be used, including "urban renewal clearings," "school yard property," and "unused right-of-ways."[76] The city's director of human resources even suggested to the mayor that a vegetable garden might be cultivated atop a city-owned building downtown. Now in the land procurement business, the Department of Human Resources hired Washington State University to perform soil analysis of various potential sites.[77] The city approved the use of land once owned by the U.S. Postal Service near downtown Seattle for the Interbay P-Patch and leased a piece of land from Safeway, a company that for the counterculture symbolized the evils of the mainstream food economy. The P-patch program was formally included in the city's capital improvements for 1974–1983, allowing the superintendent of parks to pursue funds from the Washington Interagency Committee for Outdoor Recreation for "further development of said project."[78] By the end of the year, the city program had a life of its own.

Agriculture in Seattle was so popular that the city made other plots available to groups and individuals outside the program.[79] In the summer of 1974 a group of residents on Capitol Hill jackhammered the cement loose from what used to be a parking lot of a nearby grocery store to make way for a garden: "Well, it was the most natural thing to do," they told the alternative news weekly the *Seattle Sun*; "Rip all that up and put the green back." The gardeners remarked that the grocery store gave "garbage to add to our compost pile, which is nice."[80] By the summer of 1974 even *Sunset Magazine* reported on such developments in Seattle. "The magazine of western living" published a two-page photo spread about community gardens in and around Seattle, focusing in particular on the efforts at the original P-patch: "P-patches are currently burgeoning all over King County," the magazine remarked.[81]

Community gardens were popular throughout the country in the 1970s, as urban-dwellers from Burlington, Vermont, to Portland, Oregon, made contact with the soil in their cities. Gardening, especially in a time of crisis—war and energy shortages—resonated with long-held American ideas about the role of nature and self-provisioning as a defense in difficult times. Like other counterculture ideas, the P-patch harmonized with Americans in a time of diminished public money as well. Seattle's coordinated experiment, however, was significant because it represented the first subtle introduction of counterculture environmentalist ideas of small-scale, organic, community-based, and ecotopian principles into the agenda of a willing city government. Its success as a program and enduring institution rested on its vision of restoration and renewal on a local human scale.

In November 1974, after the first citywide P-patch program was up and running and the year's harvest was in, the urban agriculture pioneers Darlyn Rundberg and PCC manager Randy Lee, along with other inspired Seattleites, attended Tilth's Ellensburg conference, where they shared their P-patch triumph with the larger regionwide ecotopian community.

Home of the Good Shepherd

By the time Tilth had founded its Seattle chapter, ideas that only a few years earlier seemed like a fantastical hippie dream—the city managing an urban agriculture program—rapidly gained favor with the municipal government and a wider public. And by the mid-1970s, environmental heretics were easy to find in Seattle. By the end of the decade, Tilth would be drawing money and support from city departments and federal grants while creating momentum from the community interest that the P-patch program stimulated. Tilth's Seattle chapter would marshal residents' evolving interest in community gardening and channel it toward larger goals of creating a steady state landscape in the city.

Carl Woestendiek would become a pivotal figure in Seattle's Tilth chapter and the success of community gardening in the city. His biography is emblematic of counterculture environmentalism's origins and Tilth's creative approach to grassroots urban renewal. Woestendiek led the effort to construct the second enduring ecotopian landscape. Like many counterculturalists of the period, he came to the city during a road trip. He stayed because he found in the neighborhoods around the University of Washington a strong community of like-minded people and a thriving counterculture scene. Seattle, he said, was "a counterculture kind of place." Before arriving in Seattle in 1973, Woestendiek had worked with emotionally disturbed children in New Jersey as a conscientious objector during the Vietnam War. After studying English at the University of Wisconsin, where he published poems in the local underground newspaper, he left for Israel to work on a kibbutz. After Israel he traveled to Auroville, a famous and influential intentional community and ashram in southern India, where he was impressed by the agriculture experiments he saw there. When he left India a year later, he made his way to Seattle and tried to put his skills as a farmer to work in the city. At the time he had no car: "I carried my tools around in a big green bag on the bus." He admitted that he "wasn't very knowledgeable" about plants back then.[82]

In 1974 Woestendiek decided to go back to school in horticulture. At the time, the best place to earn a horticulture degree in the Pacific Northwest was Washington State University (WSU). As an older student at WSU (he was thirty), he had some leeway to craft his own program, which included a creative final research project. During his time at WSU he became "intrigued with urban agriculture," though there was no official program. So his final project focused specifically on urban agriculture projects on the West Coast, and as part of the research he traveled south to tour and photograph various urban agriculture projects in California and Oregon. Soon after he returned from his tour, he began working as an intern for the extension service of WSU, a land-grant university. In 1976 the internship took him on a fateful trip to visit the Pragtree farm in Arlington, then the headquarters of the Tilth movement. At Pragtree, he said, he was "totally intrigued and the place was wonderful." The experimental farm—like New Alchemy and other groups at the time—was experimenting with parabolic greenhouses, raising tilapia, and using solar buffers. "It was beautiful," he said; "I was taken with the place." The farm was a meeting place and a kind of bioregional laboratory where the founders of the Ecotope group conducted "solar experiments" and where the managers of Seattle's larger co-ops met to exchange ideas. Eventually Woestendiek ended up apprenticing at Pragtree for three months. Around the same period (1976–1977), Woestendiek spent two summers as an intern for the King County Cooperative Extension, where he worked to preserve dwindling farmland in the Kent Valley and elsewhere. "I got to hear firsthand about the problem of preserving farmland," he recalled.[83]

Having grown up in Chicago, Milwaukee, and other cities, Woestendiek recalls, he was drawn back to work in urban areas, because, he felt, "it was really important for urban people to have some experience of raising their own food in order to appreciate what the farmers were doing for them." He believed that the "major part of the mission of Seattle Tilth" was "to make that linkage": "I think everyone in the Tilth larger network felt that way, because I know people felt good when the Seattle chapter formed and it was a vigorous chapter and that information needed to get to urban people in particular because they were making all the buying decisions." Upon his return, Woestendiek bought an old house on a large lot in Seattle's Wallingford District—just west of the University District—for $70,000. Wallingford, according to Woestendiek, had "probably the most activist community council at the time." In the mid-1970s the council was busy fighting to save an early twentieth-century school for young women, the Home of the Good Shepherd, from the wrecking ball, in another iteration of the proliferating preservation and neighborhood fights of the period. The historically significant building interested preservation groups and housed various arts organizations and political groups, a quintessential 1970s urban organizing space. But for some time, what Woestendiek had noticed about the compound were the grounds around it, especially the "tons of fruit trees, and raspberry plantings." He then got a close look at the grounds when he became groundskeeper at this urban oasis.[84]

In a serendipitous turn of events, Woestendiek found himself at one of the Wallingford community meetings in 1977, soon after Tilth's first regional meeting. The Wallingford neighborhood group had asked him to replace another presenter on a program about the endangered center's history. Woestendiek explained the idea for the center to an overflow Wallingford Community Council meeting that October, and he grabbed the opportunity to push his idea for making its grounds into an urban horticulture demonstration project. Using slides and drawings from his road trip, he convinced his auditors to support the idea, and soon they became "vocal advocates." The center could create "a model urban agriculture center."[85] Woestendiek proposed preserving the orchards on the property. He urged the city to use the site as "an educational and research facility for growing food in the city," an explicitly educational demonstration project.[86] Charting the center's future evolution, Woestendiek envisioned building a solar greenhouse on the site. In a piece he wrote for the Tilth newsletter, Woestendiek explained, "There are exciting projects happening in all our cities as people realize that life in an urban environment can be more than a consumer experience"; "What people learn by 'growing their own' will also benefit the farmer, whose exertions and skills are of-

The greenhouse in the Tilth demonstration garden with the Home of the Good Shepherd in the background, 2010. Photograph by Jeffrey Sanders.

ten underestimated by shoppers only aware of rising prices at the supermarket."[87] The activist suggested, "Blurring . . . the physical and psychological borders of urban and rural life will do us all a lot of good."[88] One neighborhood resident wrote to a city official after the meeting, saying, "I am supportive of what seems to be a trend towards more food production in the city and believe this should be encouraged by having a place where people can go and learn about agriculture in the city and more importantly, see it happening."[89] Few in the community opposed the center.

The first Seattle Tilth newsletter repeated the larger group's goals, focused on the urban environment. The group would encourage "people to grow as much food as possible in all the green pockets and crannies of [the] sprawling city." In its statement of purpose, the Seattle chapter set out broad goals that included "the creation of agriculture communities in

Seattle and the development of urban food production and exchange systems." And it set more ambitious efforts, such as encouraging the "city to plant fruit and nut trees along streets," creating "agriculture neighborhood support groups" and a "support group to promote and create local agriculture in King County and Western Washington," and developing "marketing support for farmers."[90]

By the time Woestendiek was taking his show on the road, the Wallingford neighborhood had succeeded in saving the large building and sprawling grounds of orchards and grass. Through the use of federal revenue sharing funds and money from Forward Thrust parks programs, the site had been preserved as a community and arts center with an attached playground. The playfield had become part of the Seattle park system. It would take some time before the city would fully acquiesce to the idea of an urban agriculture research and demonstration project in the middle of a new park, however. Woestendiek recalled that Walt Hundley, by then the head of the Seattle parks department, "was not intrigued." "Urban agriculture was a pretty far-out thing in 1977 and 1978," Woestendiek recalled, and "he was a fairly conservative guy." The parks department "just saw a bunch of hippies that wouldn't follow through with anything and they would be left with a big liability on park land."[91] But instead, the grounds became a success in a neighborhood that was quite receptive to the idea. At one point, Woestendiek recalls, the director of community development even showed up for a work party in bib overalls, perhaps signaling the shift in urban priorities by the late 1970s. The head of the P-patch program in the 1970s (a Tilth member) and the city's director of human resources determined that a community garden at the center "would not interfere with the rest of the park development."[92]

Before 1978 Tilth had no home base for its activities in Seattle, but after Woestendiek's presentation, the possibility of a site at the Good Shepherd building appeared strong. Woestendiek and other members of the Seattle chapter devised the blueprint for their experimental microcosm. Throughout the fall and winter of 1977, Woestendiek trundled his slide projector and easel to the downtown city library and community centers, holding meetings with interested gardeners and the general public. To the library he brought sketches of the "projected urban agriculture center," including depictions of solar greenhouses and composting units. The images, he said, added "a bit of earthy science fiction" to the meetings.[93]

Throughout the late 1970s and into the 1980s, Tilth obtained several grants from the city and federal agencies—including the Department of Energy and community block grant programs—to create a solar greenhouse, to demonstrate cold-frame and cloture gardening techniques, and most successful of all, to construct a composting demonstration center. By the early 1980s Woestendiek was in charge the composting program, with an office, a staff, and a hotline to help Seattle's gardeners that is still working today. Each year for the last twenty years, he has trained at least twenty people to become master composters, helping to build up the backyard soils of the city and making Seattle a pioneer in composting. The grants and ongoing city jobs provided a lasting funding stream for Seattle Tilth, making it one of the enduring Tilth chapters in the Northwest.

With the new experimental and educational site in the city, the group gave organic gardening tips to interested Seattleites and helped to spread the word of "socially equitable agriculture in the Pacific Northwest," particularly Seattle and its outlying areas.[94] Through the center and a series of

Tilth demonstration garden on the grounds of the Home of the Good Shepherd, 2010. Photography by Jeffrey Sanders.

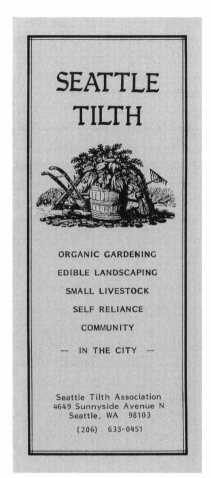

Seattle Tilth pamphlet, 1978. Tilth Association Records, Cage 602. Courtesy Manuscripts, Archives, and Special Collections, Washington State University, Washington State University Libraries, Pullman, WA.

"urban agricultural" workshops at the Pike Place Market and the Wallingford center, Tilth helped Seattleites hone their organic gardening skills and "learn more self-reliant ways of providing their food."[95] Typical workshops featured tips on beekeeping, composting, building and using solar greenhouses, and organizing community gardens, as well as talks by such urban self-reliance luminaries as Bill and Helga Olkowski, the founders of the Integral Urban House in Berkeley and authors of *The City People's Guide to Raising Food.* The Wallingford center also became home to Ecotope, a Seattle group dedicated to the "development and demonstration of solar energy

163

and conservation technologies" and their application in the Puget Sound area.[96] By the late 1970s and early 1980s, Seattle Tilth had helped to cultivate an urban constituency and consumer base that supported the cooperative community within the city and small producers without. According to Carl Woestendiek, Tilth had made worm bins and compost piles the hot topic of conversations at neighborhood potlucks and parties.

Yet in Wallingford the politics of green consumption were often practiced as a private affair. The demonstration gardens provided a public context for learning about alternative agriculture and food economies, but in an increasingly middle-class neighborhood of detached single-family dwellings, they also encouraged cultivation of private plots and a self-sufficiency that easily translated into private experience rather than a public ritual in public space. These same ideas, however, would play out differently on the other side of Seattle.

The Danny Woo Community Garden

While Rundberg and Woestendiek created gardens and demonstration centers in Seattle's white, middle-class north end neighborhoods, the longtime neighborhood activist Bob Santos was busy planning an ambitious vegetable garden of his own in the International District. Seattle's "Chinatown," the area was a part of the original Model Cities neighborhood during the mid-1960s and the site of intense organizing by the early 1970s. In 1975 Santos sat down with Danny Woo, a business owner with deep family roots in the district and the owner of the Quong Tuck Bar and Grill, on King Street, a local watering hole for neighborhood activists. Woo and his family owned a stretch of vacant land on a steep slope next to the I-5 freeway, which had cut across the edge of the neighborhood in the early 1960s. The vacant lot sat near the center of an area that was once the thriving core of Seattle's *Nihonmachi,* or Japantown. In his meeting with Woo, Santos hoped to negotiate a lease for this rubble-strewn open space full of blackberry brambles and tall grasses. His intention was to use the hillside as an important food source and open space for seniors, many of them former farmers, who lived nearby in low-income housing. In his memoir, Santos recalls that when he proposed a price of one dollar for the site, Woo nearly "swallowed his pipe."[97] Santos was finally able to convince Woo to rent the property to his neighborhood organization, but with no promise of a long-term lease.

Santos knew how to get things done in the neighborhood. A Filipino American native of Seattle active in a community that historically included most of the city's Japanese, Chinese, and Native American citizens, Santos learned early to navigate neighborhood politics and multiethnic coalition building. The International District also sat beside the Central District, making the area perhaps the most ethnically diverse part of Seattle at the time.[98] An early leader in the fight for racial justice in the mid-1960s, Santos had taken the lead in Seattle's Asian American empowerment movement, which emerged alongside similar empowerment movements among the Latinos, African Americans, and Native Americans who began to assert themselves and the idea of ethnic pride as part of their politics. Santos was also part of the influential "gang of four," a quartet of 1960s civil rights leaders (the others were Roberto Maestas, of El Centro de la Raza; Larry Gossett, of the Central Area Motivation Project; and Bernie Whitebear, of United Indians of All Tribes) who, according to Santos, had found common cause by proximity. They shared not only a neighborhood but also a meeting space at St. Peter Clavers, a church in the Central District. Throughout the 1960s, the church had been the center of feverish organizing and marathon consensus-building meetings among Seattle's civil rights community, Black Panther Party, and others. Santos recalls that each of the gang of four could count on the others to turn out their own constituents—the same dependable four hundred protesters, he recalled—at a moment's notice for any direct action they planned. Santos and many of his fellow Asian activists, for instance, picketed with the UIAT outside the gates of Fort Lawton.[99] By the early 1970s, Santos could claim battle scars and great successes, as well as the support of mainstream businesspeople in the International District.[100]

A self-described "urban warrior" of 1960s civil rights and community politics, Santos did not fit the typical image of a pioneer in the urban environmental movement.[101] In popular culture, contemporary discourse about the nature and history of organic foods and urban gardeners focuses on images of privileged, white, middle-class consumers, Whole Foods shoppers, or even stereotypical hippies.[102] To a certain degree Tilth organizers (e.g., Musick and Woestendiek) fit this conception. Yet as Tilth moved to the west side of the state, it built networks of cooperative food stores and buying clubs, organic gardeners and composters, and appropriate technology enthusiasts, forming a crucial infrastructure of tools and ideas that

Bob Santos, 1971. Photograph by Phil. H. Webber. *Seattle Post-Intelligencer* Collection, image no. 1986.5.54102.1. Courtesy Museum of History and Industry.

could be applied and used by any of Seattle's various neighborhood revitalization projects. Investigating how these tools and ideas were used, mobilized, and eventually institutionalized reveals a less-familiar story of grassroots urban environmentalism. This is where Santos comes in.

After expanding the Model Cities program with its "planned variations" program, the Nixon administration canceled it in 1974, ending the flow of money that had helped to subsidize many important local experiments in neighborhood-based urban renewal and urban environmentalism. Yet before the program faded from the scene, to be replaced with the stingier urban policies of the late 1970s, neighborhood activists used federal funds to create Inter*Im, the International District Improvement Association. The organization had its origins in the Jackson Street Community Council, a neighborhood organizing group begun not long after World War II, but in the 1970s activists in the International District were able to obtain start-up funds for the newer organization. Santos was recruited in 1971 to run Inter*Im and elaborate on the goals of Model Cities, providing community health services, recreation facilities, housing rehabilitation, and especially outreach work with the area's many elderly residents. In those first years, Model Cities was "a source [of funding] that kept the operation of Inter*Im alive," Santos explained.[103]

During this period of heightened community organizing in the International District and in the nearby Central District, the city drew up plans to build a large sports stadium called the Kingdome in the neighborhood just south of Pioneer Square and on top of low-income housing where some of the neighborhood's elderly Filipino residents lived. The ensuing fight between neighborhood activists and the city would further politicize the Asian American movement in Seattle and mobilize the neighborhood around the control and production of space in the area. Referring to the rapidly coalescing neighborhood pride and Asian Power activism, Santos and others picketed the sports stadium's groundbreaking events with signs that read "Hum Bows, Not Hot Dogs," and they pummeled the mayor and city officials with mud balls to show their disapproval for the plan.[104] At the Kingdome protests, according to Santos, "Bernie and Maestas sent the troops down" to help in the community's efforts to assert control over development in the neighborhood.[105]

From its inception, Inter*Im served very clear goals that set it apart from the work of activists in the Central District or Wallingford and those who fought for the Pike Place Market. Santos and the others saw their preservation and urban farming efforts not as an aid to gentrification but as active resistance to it, their principal goal being to prevent their neighborhood from succumbing to outside commercial influences, as many other West Coast Chinatowns already had. At the same time, Santos and Inter*Im shared the tools and attitudes of Model Cities and market preservationists. During the last year of Model Cities funding, in 1973, International District residents battled developers and pressured the U.S. Department of Housing and Urban Development (HUD) to spend money on low-income housing in the neighborhood. "Our staff went to visit HUD almost weekly," Santos recalled. During the Kingdome conflict, the Model Cities activist Diana Bower, who had been active in MC physical environment programs, took a job coordinating the neighborhood and the city regarding the Kingdome's impact on the area. Working closely with the

Inter*Im staff, she helped young activists to organize senior citizens there, especially around the need for better housing.

It was during this period, after weeks of interviewing seniors, that Bower proposed the idea of a garden for the neighborhood. In his memoirs, Santos calls Bower the "compassionate consultant" who, "after talking with a number of International District residents about their needs," had come "to the conclusion that after meeting housing, health care, and nutritional needs, residents wanted the ability to garden and raise their own vegetables."[106] Soon after, the Inter*Im board began searching for a garden site, and Santos found himself clearing car tires from the hillside's blackberry brambles and spending three hours taking soil samples with Bower. In the years before the community came together to tend the land, it was "the repository of excess materials from construction of Interstate 5." The "rubble and clay" of the hillside had to be stabilized with a deeply anchored retaining wall at the bottom of the slope. The land was located behind the Seattle Housing Authority's Turnkey Housing Project, a site that a soil engineer's report described as consisting of "grass and blackberry bushes" in "brown sandy gravelly silt with rubble and organic debris."[107]

In 1975 Bower reached out to her friend Darlyn Rundberg for help in creating the garden. According to Santos, "Darlyn jumped at the idea of working with the I.D. activists and residents to build a community garden." Eventually with support from Inter*Im and the Seattle Chinatown I.D. Preservation and Development Authority and with financial assistance from the Hunger Action Center, the project began. Over 180 volunteers assembled to construct the terraced garden plots. According to Santos, Bernie Whitebear sent his people down from Daybreak Star at Fort Lawton to help with the project. Elaine Ko, a young Inter*Im community organizer, remembers that everyone was called on to do his or her share in preparing the terraced garden. The massive undertaking brought a range of political activists and organizers from the old Model Cities neighborhood to heave 750 railroad ties into place. "On a typical Saturday," according to Santos, "the hillside was filled with hard working, sweat-soaked bodies."[108] In addition, the soil had to be intensively enriched with 350 cubic yards of horse manure from Longacres Racetrack, south of Seattle. It will be "some time before the Tilth of the soil is ideal," advised Tilth's newsletter.[109] With the additional help of HUD grants and donated property, the group of urban

Building the Danny Woo Garden, 1977. Tilth Association Records, Cage 602. Courtesy Manuscripts, Archives, and Special Collections, Washington State University, Washington State University Libraries, Pullman, WA.

warriors turned farmers transformed the wild, overgrown hillside into a beautiful terraced garden. Walt Hundley, who had resisted Woestendiek's efforts at the Wallingford center, was quite supportive of the gardening efforts in the International District, according to Santos. Having known Hundley from Model Cities, Santos believed that "he was very familiar" with their goals and that he had told his staff, "Don't bother with that kind of stuff. They're doing a good job. Let's work with them . . . give them a ninety-nine-year lease on adjacent city parkland to Inter*Im."[110]

From Tilth's point of view the coordinated effort exemplified its expanding commitment to "create an ecologically sound and socially just

The Danny Woo International District Community Garden, facing west, 1978. Tilth Association Records, Cage 602. Courtesy Manuscripts, Archives, and Special Collections, Washington State University, Washington State University Libraries, Pullman, WA.

agricultural system."[111] Tilth described itself in 1975 as an "educational and scientific organization with an emphasis on agriculture as the heart of a systematic transition to an equilibrium culture in the Pacific Northwest."[112] With the P-patch system up and running by middecade and with Tilth's ability to secure grants and articulate its vision to the public, Rundberg and other Seattle activists capitalized on the momentum and interest in the program. Using the city program as a foundation from which to perform experiments in food production and energy, Tilth's ecotopians exposed the public to alternative land use, organic gardening, and alternative energy. Making this transition visible and demonstrable in the city was Rundberg's particular talent. Emphasizing the ethnic composition of the community of elderly Asian citizens whom the garden served, Tilth's newsletter stated, "On this day, people of the community reclaimed a bit of soil for themselves in a garden built with some of the railroad ties their forefathers had placed during construction of the country's railroads."

The garden plots' contents also reflected the district's cultural makeup. The plots were planted with "ethnic vegetables," such as Chinese cabbage, bitter melon, bok choy, Japanese eggplant, and Chinese parsley. After the garden's first season ended, Tilth said it was "fast becoming a horticultural delight," one that had "enhanced the spirit of a community" that could "come to articulate its problems, then persistently pursue solutions."[113] On the Seattle hillside, ecotopians attempted to link culture and landscape.

Tilth may have understood the effort as a triumph of ecotopian principles, but Santos saw it slightly differently. He described his faction as "urban warriors," whereas as Tilth members were "environmentalists": "We didn't even know those terms then. We just knew that if we built a garden for the old folks, they would be out in the fresh air, and we would provide exercise because they'd have to climb the hill to get up there. And it was cost savings to them, so we were looking at it from a different angle." And yet those other identities were in the air. Santos explains, "We kept hearing that the I.D. garden was one of the very few green spaces and open spaces in the urban center of downtown Seattle." He added, "We started to get more interested in what we were doing, and so we aligned with . . . Tilth"; by 1978 "they sent people down to teach our gardeners how to compost and how to build the compost bins and that's when we started to get involved in this ecofriendly kind of stuff." To the extent that they were aware of it, the transition was gradual: "We were interested in housing and health and trying to preserve that neighborhood so that the old folks could survive, and then later on we were building housing for the workforce and then building housing now for the families. . . . The garden," he says, "is sort of central to everything we do now. Maybe unconsciously we are very involved in the environmental movement; we preserved that area, not a greenbelt, but its undevelopable land."[114]

Santos and the International District garden contribute to and illuminate the story of these diverse early steps toward the sustainability movement. The Danny Woo Community Garden would eventually take shape as an important symbolic, political, and environmental expression. Santos explained that the garden combined the nutritional, aesthetic, economic, and political; it was a place to which the community and leaders could proudly point as "an example of what can happen with strong community involvement."[115] Perhaps as important, the terraced garden near the free-

Kobe Terrace, Danny Woo International District Community Garden, facing east, Sept. 13, 2004. Photograph by Beth Somerfield. Seattle Parks and Recreation, Department of Parks and Recreation Photographs, Photograph Collection, item no. 150150. Courtesy Municipal Archives.

way represented a confluence of forces that were working together at mid-decade in the city and that were rooted in the early 1960s atmosphere of neighborhood organizing and 1970s ecotopian visions.

The Danny Woo Garden and Santos's story helped to explain the cross-town marriage of these activist strains in the city while challenging assumptions about the main constituencies, dominant concerns, and cast of actors in the environmental or local food movement. Each of these community agriculture and food projects show how ecotopian ideas could work for different groups and different political goals in the city. In Wallingford Tilth served to revivify an already mobilized white, middle-class neighborhood while unintentionally reinforcing a more private green politics and mode of consumption. In the International District, however, and in the P-patch program throughout the city, urban agriculture offered both

172

a green symbol of resistance to gentrification and a source of cultural and community identity and empowerment.

The massive project of making a rubble-strewn urban hillside next to a freeway into a garden served as a powerful lesson for Seattleites in general. Tilth's role in making this landscape thus served multiple purposes. Certainly the garden presented an example of an urban answer to the New Alchemist call to create beautiful and productive landscapes to influence the larger culture. But it also worked on the consciousnesses of urban consumers estranged from their own food supply. The Tilth newsletter's description of the garden and the huge efforts to transform the hillside revealed the didactic and symbolic quality of Tilth's larger project. "What is strange," wrote Musick, "is that we, in this country blessed with such fine soil, are now finding it necessary to go to great length to preserve our prime agricultural soils from inappropriate uses." Describing Tilth's work in the International District, he said, "How ironic that we struggle here with our clay hillside to provide food-producing soil, while around us we watch the demise of agriculture on the prime soils of our nearby valleys."[116] For Tilth the dispersed neighborhood gardens, the terraced hillside project, and the urban agriculture center served as counterspaces, tangible and functioning alternatives to the preceding twenty years of metropolitan changes and the simultaneous destruction of rural food-producing landscapes on the city's periphery.

The three counterspaces that Tilth helped to create served as models that ultimately shaped lasting changes for Seattleites' expectations of urbanism. If the Pike Place Market symbolized a lost connection to nature in the early 1960s, by 1970 it showed signs of life. In 1974 the market, a symbol of Seattle's urban renaissance and a newly defined historic district, thus found its identity tethered to the fate of small farming in the region and nearby metropolitan farms. Produce, or food, was the energy and soul of the market and had become an important symbol of urban and regional restoration and renewal, part of Seattle's new image. Tilth's organizational efforts and the loose ecotopian network that it marshaled provided the infrastructure, markets, distribution, experience, and educational mission that complemented this urban and rural rebirth. Tilth's activities had helped to nurture this shift in consumer sensibility, and its various enterprises would continue to support it.

As early as 1971, a King County Environmental Development Commission report on open space recommended that the county "preserve prime agricultural lands and significant other farmlands in the open space system."[117] Though still couched well within the late-1960s paradigm of recreation and broadly defined open space, the report detailed the rapid loss of rich soil in the areas around Seattle and set out criteria and policies for its preservation through zoning changes and tax relief for farmers.[118] Following these initial expressions of concern for farmland preservation, King County voters passed a series of ordinances and action programs during the early 1970s, going so far as to create a moratorium on further urban development in farmland areas in 1975 until a thorough study could be performed.[119] The county had begun to respond to increased public pressures regarding open space preservation and the loss of nearby farmland. While activists and government officials worked to preserve farmland, Seattle's Department of Community Development grappled with the other side of this decline in agriculture, producing the "Pike Place Market Farmer/Vendor Study," an examination of the market's "history and traditions," which the agency believed were "being threatened." It sought to ascertain "the possible means of attracting additional farmers to the Market."[120] With the work of Victor Steinbrueck, the market became a symbol of the city's new organic ethos, which included historic preservation, recycling, and restoration and an affirmation of the past. Its funky old produce stalls and fragrant corridors evoked a lost relationship between the city and its hinterland, as well as the possibility of regaining a more "organic city."[121] At the market, the ecotopian linkage of city to hinterland could be made manifest.

The city's report detailed the market's past glory and recent rapid decline, a story of lost farmland that ecotopians and market preservationists knew well. Having made an effort to preserve the failing turn-of-the-century farmers' market, the city faced a problem: few farmers and even fewer farms. In 1969, according to the report, the number of vendor permits for farmers had reached an all-time low of 42. Just before the war permits had reached 515, allowing farmers to sell their fresh produce directly to the consumer at the market in the late 1930s. The 1974 report delved briefly into the larger societal problems of modernization and industrialization's effects on agriculture and finally argued in an ecotopian vein: "If we are to attract more farmers to the Pike Place Market, we must determine how the Market as a microcosm fits within the agricultural macrocosm, and ensure its continual position there."[122]

The report drew some interesting conclusions about how to run a farmers' market without farmers or farms, suggesting among other things that perhaps larger-scale farmers could tailor their operations "specifically for sale of products at the Pike Place Market" or that the city might subsidize local farmers and "provide agricultural land, equipment and technical resources to persons who would commit to selling their products" there. The report's authors even suggested that the city encourage "home gardening and pea-patches and attract these hobbyists to sell their products at the market."[123] The report concluded that "for any of the programs to work, or even for the existing farmers to be able to continue their activities at the Market, there must be consumers" and that "any effort taken to reinforce the role of the farmer at the Pike Place Market must be complemented with an equally strong effort to attract more buying customers."[124]

Tilth had out of necessity begun "liberating" land through trusts, communal or public ownership—as with Pragtree farm—and other creative means of funding. Tilth managed to farm land and create markets outside or alongside the traditional "Safeway" food economy, thus demonstrating an ability to make things happen despite the slow movements of city and county government. In 1974 those associated with the group officially founded the Evergreen Land Trust (ELT), "a non-profit corporation established to remove land from the speculative market and make it available for community and environmental purposes."[125] The trust's properties included Pragtree farm, which served as the headquarters of Tilth. The group also ran communal houses in Seattle; the River Farm, in Whatcom County; and Walker Creek property in the Skagit Valley.[126] Such land preservation, trust, and grant funding ideas had been the subject of panel discussions at the Ellensburg conference, in *Seriatim*, and in the pages of Tilth's newsletter throughout the decade.

By the late 1970s Tilth was adept at using HUD, the Community Services Administration, and block grants to make its projects happen. Tilth also reported to its membership on statewide and countywide grassroots efforts of "farmers and environmentalists in the Puget Sound basin to prevent further losses of prime agricultural lands."[127] As the city and county became more committed to urban and rural renewal based on agriculture and the Pike Place Market, members of Tilth would contribute well-honed tools for assisting in this transition. Darlyn Rundberg, for instance, worked during the late 1970s for the new King County Office of Agriculture, where she developed an internship program "to train new farmers as part of King County's effort to preserve local agriculture."[128] In the early 1980s these projects would begin to bear fruit. One of the more innovative projects undertaken by this three-way marriage of counterculture environmentalism, county land preservation, and the city's push to restore agriculture's cultural and historical role found its impetus in a surprising source, but one that exemplified the melding of social justice and ecotopian ideals.

After the Vietnam War, Seattle social service agencies were overwhelmed by relocated refugees. Washington State had the nation's sixth-highest Indochinese refugee population, including Vietnamese, Laotians, Cambodians, and Hmong, with over 7,097 living there by 1979.[129] Many of these people were formerly farmers. In an effort to create opportunities for the refugees, who were overwhelming public assistance programs in Seattle, the city and county created the Indochinese Farm Project in 1982. The farm, located on King County–owned land, served mainly Hmong and Mien people. The nonprofit operation, sponsored by the Washington State Commission on Asian American Affairs, King County Cooperative Extension Program, Pike Place Market Preservation and Development Authority, and a local church, provided the refugees training and educational opportunities. According to the *Seattle Post-Intelligencer*, the program provided "a means for them to become self-supporting as truck farmers."[130] As the urban renaissance of Seattle took hold in the mid-1980s, and as capital and population returned to the city, these refugees farming in Woodinville, near Seattle, would become one of the major success stories of ecotopian efforts to renew agriculture in the region and to bring local agriculture back to the urban fabric. The Hmong sold their produce at the Pike Place Market, to the inner-city co-ops, and to several high-end restaurants, including a French restaurant in Wallingford that, according to the *Post-Intelligencer*, bought "its fresh produce year round from the refugee farmers."[131] The Hmong exemplified the merging of public funding for agriculture with the cultural significance that Tilth imputed to it.

Ten years after the Spokane world's fair and the Ellensburg conference, food had made the transition from the means of an environmental revolution to a fetish for young urban professionals returning to the city. Recovering from the steady population decline after 1960 and the Boeing bust of the early 1970s, the city began to regain population, reaching levels close to those of 1962. The city that ecotopians remade now lured people and investment. Fresh local produce, the Pike Place Market, and a burgeoning restaurant scene seemed to symbolize the city's rebirth. "Thanks to the new-found abundance of fresh local ingredients and the popularity of what is fast becoming a uniquely American style of cooking," wrote a *New York Times* reporter, "the Northwest is now enjoying a food renaissance and this lush green city is the focal point."[132] Of course, this lush, green city and its connection to an "expanding network of small farmers" would not have existed without the labors of Tilth and people such as Musick, Woestendiek, Rundberg, and Santos.

The national coverage confirmed what Seattleites already knew about their changing city. The *Seattle Weekly*—the main chronicle of Seattle's post-1970s reinvention and a guide to the finer things in Seattle's new consumer identity—had spotted the food revolution in Seattle several months before the *New York Times* had. As Seattle's unofficial booster organization, the *Seattle Weekly* effectively sold Seattle's rebirth to visitors and residents. The same publishing company also produced *Northwest Best Places* and *Seattle Best Places*, regional guidebooks that reviewed, recommended, and promoted the local scene. Musick and his partner on the farm graced the cover of the February issue of the *Weekly*, which announced, "Local producers are transforming the way we eat."[133] The related article, which focused on food, claimed that five years earlier "anyone would have thought he [Musick] was crazy" for his wild salad ideas. But, it claimed, "part of Pragtree Farm's success has been the good luck to grow and change at the same pace as the Seattle culinary market."[134]

This was one way to look at it. Certainly Tilth and ecotopian experimenters of the early 1970s had accommodated to the shifting sensibilities of the urban market, but they had also effectively created a market for those tastes through their persistent promotion and consumer education and through preservation and restoration of agriculture in the city and on its suburban edges. The lush green city was born not from suddenly awakened consumer desire but from the steady consumer activism of many years. If something approximating a "steady state" had been reached—the emergence of a regional cuisine and culture of nearby food—then ecotopians could take much of the credit. Of course, Musick and his group had done much more.[135]

Key to this urban transformation was the Pike Place Market, which was able to survive as a symbol of Seattle's new food culture and consumer ethos in part because it had something to sell.[136] Tilth continued to encourage novel approaches to nearby food production, including "urban forests," "hedgerow habitats," and Musick's wild greens. Affiliated Tilth projects had transformed an industry that began to explode by the late 1980s. Writing in 1991 for *In Business Magazine*, Musick detailed the work of his longtime friend and Tilth founding member Gene Kahn, whose "New Cascadian Survival and Reclamation Project," an early 1970s organic farm in the Skagit Valley, north of Seattle, had metamorphosed into a multimillion-dollar business. With the help of Woody Deryckx, another Tilth founder and an attendee at the Ellensburg conference, Kahn had begun the process of selling his successful operation, now named "Cascadian Farm," to Welch's in 1990. A leader in the growing market for processed organic jams, frozen potatoes, and other vegetables, Cascadian Farm was an impressive success or a sell-out, depending on how you looked at it. Certainly Cascadian Farm had grown far beyond Pragtree and the small-is-beautiful goals of the movement, but Kahn had also helped to create a booming market and sensibility for alternative foods and for the vegetables he procured from the farmers he called his "50 certified organic farms" in the ecotopian region and far beyond.[137] An ethos invented at the intersection of Pacific Northwest consumers and the Skagit Valley counterculture had gone global.

A dependable system of certified organic farms and produce was itself an outcome of Tilth's ecotopian work. Oregon Tilth (which broke off from Washington in the 1980s) became the leader in organic certification before the USDA got into the act. Once seen only on obscure brands of tofu and beets grown in the Northwest, Tilth certification became a dependable indicator of food items' organic bona fides. By the late 1990s Frito-Lay, certainly a mainstream company, had begun to use Oregon Tilth's certification stamp for its line of organic potato chips. Those critical of this shift toward mass marketing, convenience, and largeness in general could be reassured by Tilth's other efforts to nurture small nearby agriculture.

A fluid organization from the start, Tilth worked on various scales at once, including those of mass consumers, regional agriculture, and cities. While Kahn processed his organic jams and vegetables grown on dispersed organic farms, the P-patch program in Seattle continued to develop its urban acreage, which by 2000 included at least twelve acres in forty-four neighborhoods throughout the city. The Puget Consumers Co-op went on to create its Farmland Fund, which began operation in the 1990s and continued to develop and preserve local agriculture for a growing network of organic food specialty stores, burgeoning neighborhood farmers' markets, and of course, the Pike Place Market itself.

In the 1980s the *New York Times* acknowledged that the demand for fresh local produce had "contributed to the revitalization of Seattle's Farmers Market, Pike Place Market."[138] By 2000 the market had reached a height of 112 farmer vending permits, almost double the number in the early 1970s. Among these new farmers crowding into once empty market stalls were the Hmong from King County. Many of the families who had begun farming in the early 1980s as part of the Indochinese farm project began selling fresh flowers at the market during the late 1980s and continue today. The market officially became the number-one tourist destination in the state of Washington by 2003.[139] Although the negative consequences of U.S. farm policy that began under Nixon had only recently accelerated during this period, by the time of the city's late-1990s dot-com boom, the market seemed a far cry from the suffering scene it was in the early 1970s, when the city was ready to close it. Through its various ecotopian projects and networks, Tilth had created the foundation of food activists that would reemerge with new vigor by the early twenty-first century.

5

Home

Jody Aliesan estimated that between the winter of 1979 and the following spring, over 4,500 strangers milled around her kitchen, examined her living room, and peered at her raised-bed gardens. At a time when the energy crisis of the 1970s was at its most severe, Aliesan made the private realm of her North Seattle home a public display as part of the U.S. Department of Energy (DOE) Appropriate Technology Small Grants Program. She offered the throng that filed through her home in the series of open houses the chance to see how someone could live "as responsibly as possible."[1] The DOE dubbed her self-sufficient house an "urban homestead," but Aliesan described it more simply. It was, she said, "how we live"; the DOE grant, she added, made it "possible to share it."[2] In her open houses Aliesan showed by example how urbanites could tread lightly on the environment (she later taught further lessons through the weekend columns she wrote for the *Seattle Times*). While the public walked through her house, Aliesan donned a beekeeper's suit and climbed a ladder to gather honey from her rooftop hives. She labored in her extensive raised-bed vegetable garden. In her weekly columns she explained how to insulate a water heater and weather-strip a house, make tofu, buy in bulk at a co-op grocery store, live without a car, recycle cans and bottles, and save urine in a chamber pot for the compost pile.[3]

In the notes scrawled out on hundreds of reply cards and thank-you letters, individuals, schoolchildren, and neighbors who toured her house expressed an overwhelmingly positive response to the spectacle of Aliesan's work. Of those visitors who took the time to comment, many were "inspired" and "amazed" that an "ordinary looking household could conserve energy."[4] Most respondents pledged to improve their own homes: "We are going total compost/recycling . . . [and] we're going to build a retrofit passive solar system this spring," one writer promised. Aliesan's way of life appealed to a Seattle public that by the late 1970s had been primed for her message of survival and self-sufficiency in difficult times.[5] One visitor wrote that the way Aliesan lived seemed "to be in about as much harmony with the Earth as is possible in an urban setting."[6] By 1979, nature, its processes and resources, had become highly tangible to environmentally aware Americans contemplating scarcity for the first time since the Great Depression. Aliesan encouraged Seattleites to trace such connections between personal consumption in the home and a seemingly distant hinterland of natural resources. Her home appeared to harmonize city living with nature, bringing the idea of "whole earth ecology" into it. Aliesan's "homestead," a blueprint for survival and self-reliance in the city, mixed familiar homespun nostalgia with a radically new version of the meaning of home in the postindustrial city.[7]

In the mid-1970s many Tilth counterculturalists, environmental heretics, and later, young professionals saw in more ecologically oriented landscapes such as Seattle's the promise of a malleable urbanism, a place where the architecture and functions of the city could work *with* natural systems.[8] Ernest Callenbach had conjured a postindustrial fantasy where people planted flowers in the cracks of sidewalks and streams gurgled through downtown streets reclaimed from the automobile.[9] His vision offered escape for some. Yet few people moving to Seattle during the 1970s had necessarily read *Ecotopia*. What many would-be ecotopians shared was the idea of redeeming and recycling the older meanings of the city, an idea that had already made its mark in Seattle's political culture and urban ethos beginning with the Pike Place Market fight in the early 1960s. With International District preservation efforts and the expansion of Model Cities to north Seattle neighborhoods, the politics of neighborhoods began to dominate urban affairs. By the mid-1970s this profoundly local and domestic thread of urban environmentalism became a focus of a more dispersed and increasingly ecology-inflected community activism. In Seattle's down-

and-out streetcar-era neighborhoods, newcomers and longtime Seattleites alike brought new energy, increasing numbers of them cultivating their own public and private ecotopias—with mixed results.[10]

Aliesan remembered finding a sense of home the moment she stepped off the train in Seattle's King Street Station at the start of the 1970s. She had reached the middle point in an arduous journey. In 1968, in response to the murder of Martin Luther King Jr., Aliesan went to Birmingham, Alabama, to help organize civil rights efforts there. When the internship that took her to the South ended, she was told, "with respect," she says, that the best thing she could do would be to work with her own people. She decided to leave her graduate studies in English at Brandeis and instead do antiwar work with the Vietnam Moratorium Committee, in Washington, D.C., then planning the moratorium and March on Washington scheduled for October and November, respectively. She soon transferred to Chicago to work for antiwar organizations there. She said, "I remember the Chicago cops sorting our trash, men with rifles on the rooftops across the street, [and] heavy wiretapping," and she mentioned that the rally she helped organize was the biggest in Chicago's history. But soon she left her Chicago-based antiwar activities, fed up with the blatant sexism in the movement. At the turning point, she recalls, she walked "out of the Moratorium office and into the Chicago Women's Liberation office." By May 1970, she and a friend from Chicago, Randy Lee, decided to move to the Northwest. She herself had never been there, but her mother's family had once farmed in the Sammamish Valley, near Seattle. When she arrived in the city that spring, she said, "I knew I was home."[11]

Aliesan's journey to Seattle traced a familiar course on the map of New Left and counterculture politics during the 1960s and the early 1970s, a political trajectory moving from civil rights, to antiwar, to second wave feminist and environmental politics. Like many of her compatriots who found refuge in Seattle or the other places worn-out activists called home after 1970, Aliesan was drawn to a life that she saw as closer to nature in a city with a progressive history.[12] Explaining why she thought so many people with histories similar to hers embraced environmental politics after the 1970s, Aliesan said: "After war, confrontations, insurrections, assassinations, brutality, economic recession, and the sheer exhaustion of pushing over into a new watershed—a desire to retreat (as in spiritual), recover, and heal. To do something constructive that was also personal, peaceful, satisfying, sensory, and beautiful."[13] In Seattle, Aliesan and her traveling companion found just the place, a city of distinct neighborhoods full of old single-family houses and an independent, political spirit. In this new community, she found that the "pace was slower," and she saw that people had "more awareness and love for the land"; the city was conducive to her values and desires as a "philosophical/lifestyle environmentalist." Though she believed that cities "were probably not necessary," she thought in Seattle that she and her friends and neighbors might make the "best of a bad situation."[14]

During the 1970s many activists had worked to erase the boundary between the personal and the political and between the private and public spheres in their politics. Like other activists who had navigated the cultural politics of gender and environmentalism during the 1970s, Aliesan surely regarded house and home with ambivalence.[15] Home, with its potent mix of ideological residue regarding gender roles, labor, and equity, sat at the center of these politics. Many Seattleites questioned received ideas that the notion of home implied: the isolation and homogeneity of "little boxes made of ticky tacky," and the environmentally destructive, farmland-eating consumption of the postwar housing industry. As part of this ongoing struggle to define a personal politics of place, Aliesan's work sought to lay bare the connections between private life and public consequences. Consumption also lay at the core of her work, as it did for much of postwar environmental politics.

For many young people in the late 1960s and early 1970s, this ambivalence about home drove them to retreat from suburb and city alike. Members of Tilth moved back to the land in an effort to connect with a more balanced experience of dwelling close to the earth.[16] Yet despite the popular narrative of hippie dropouts and communes in the 1960s, many activists remained and labored hard to remake the city and neighborhoods. No matter how radical, they still embraced home's nourishing associations in their efforts to dwell more benignly in the city: growing their own vegetables, taking responsibility for their wastes, and experimenting with new definitions of family and community. That these changes emanated from the putatively private domestic realm made them all the more interesting for their implications concerning the evolving notions of public and private politics and space during the 1970s. By the beginning of the Reagan era, privatization and private versus public space would take on far more

sinister meanings for many people who, as did Jody Aliesan, sought to re-
structure the social order of American society while laying bare the often
obscured relationships between city and hinterland. With prodding from
activists such as Aliesan and with the example of neighborhoods such as
Fremont, Seattleites began to question the sets of social and environmental relations that older ideas of home and community implied.[17]

Conversely, the new idea of home emerging in Seattle, including gentrifying neighborhoods, such as Wallingford and Fremont, and the sense
of private political virtue and even refuge that it could foster also helped
reinforce traditional conceptions of the home as a private realm of self-
sufficiency and consumption isolated from a public sphere. To an extent
these activists, the avant-garde of Seattle's counterculture gentrification,
echoed suburban flight in their intense localism. Counterculture activists
who sought to remake domestic spaces one house and one neighborhood
at a time walked a tenuous line, then, between the potential for a new public politics of urban environmentalism and a virtuous private retreat from
the public sphere.

Making the "Middle City"

Newcomers and longtime Seattleites with ecotopian dreams could tap into
the flourishing local and West Coast counterculture movement for an image of liberated domesticity and green consumption. Creating an alternative conception of the home in which the domestic sphere of private
choices had obvious public consequences preoccupied this movement.
Counterculturalists who may have rejected the suburban and consumer
ideals of their parents' generation would attempt to rescue the domestic
realm from its older associations, yet like their 1950s counterparts, the
1960s counterculturists had their own magazines and models in this up-
dated "kitchen debate"—although those of the latter suggested ways for
merging ecology concerns with urban neighborhood politics, with a politicized domestic consumption as a central feature. The resources available
in the *Whole Earth Catalog* (*WEC*), for instance, offered the most influential
source of these ideas. Each issue of the *WEC* exposed readers to this do-
it-yourself kind of "ecology," which linked personal choices to a broader
movement for change and survival.[18] For example, from "Ecology Action,"
an environmental advocacy group located in Palo Alto and San Francisco,
the catalog sampled neighborhood-empowering ideas such as the "Life-

house Concept": "a neighborhood community center, set up in your own home, where help and information on the starting and running of food co-ops, gardens, and compost heaps, recycling of garbage, and planting of sidewalk trees (among many other things) can be found. . . . The lifehouses will provide an opportunity for all of us to be involved in exercising our will to survive."[19] In featured advertisements and lengthy lists of recommended reading, the catalog connected its readers to the flood of Earth Day–inspired pulp "ecology" paperbacks with titles such as *Earth Tool Kit/ Environmental Action* (1970), *The Environmental Handbook* (1970), and *Ecotactics* (1970).

The *Environmental Handbook,* for example, addressed everyday practical action and responsibility in short how-to essays on recycling and composting but also published more lofty excerpts of works by Lewis Mumford and other intellectual heroes of the movement. But one 1968 essay, "The Recovery of Cities," by the group Berkeley People's Architecture (PA), notably drew a connection between ecology and transformations in the built environment of the urban home. People's Architecture worked in Berkeley around the time of the People's Park events, and its agenda reflected the blend of concerns that animated many West Coast counterculture environmentalists. People's Park activists argued that "the world of power, politics, and the institutional shape of American society" and the "world of ecology, conservation and the biological shape of our environment . . . are no longer separate or separable issues."[20] Like the later and more famous Farallones Institute, People's Architecture worked to remake the urban landscape in the posturban renewal era. Five years before Ernest Callenbach popularized the idea of an ecotopia, the Berkeley group had begun to map out an urban vision of it, one that sounded fantastic but ultimately came close to the practical ideas that Aliesan and many others, including Tilth, would interpret for Seattleites.

Drawing implicitly from the War on Poverty and the Model Cities program, People's Architecture suggested a wholesale redesign of cities and the decentralization of power in neighborhood groups. Theirs was a persuasive vision of gentrification, but dressed in the guise of ecology. People's Architecture blamed the imperial control of the old downtown power structure of financiers and law firms for laying waste to older neighborhoods, spreading suburbs into farmland, and building ever-taller office buildings. By attempting to merge ecology with a concern for the built environment, PA took this vision a step further to suggest dismantling existing high-rises and reusing their materials; it advocated closing off downtown streets and using them "for orchards, vegetable gardens, [and] parks"; and it called for the elimination of private automobiles. The group even suggested taking advantage of southern exposures in office buildings for "hydroponic gardens."[21] People's Architecture cared most about what it referred to as the "middle city," arguing that "the downtown environment will be reclaimed when the middle city defeats the bid of downtown for its territory and when suburbia becomes self-supporting communities." By *middle city* these activists meant neighborhoods with older housing stock and ghettoes (which were, they argued, "in many ways . . . the most together communities"). Here counterculture environmentalist rhetoric echoed concerns of Steinbrueck and other critics of mid-1960s urban renewal, especially the Model Cities activists: "Paradoxically, the oldest and poorest sections of the city have a head start," PA argued, maintaining that such neighborhoods would "be the first areas to show the rest of the urban population the way to an ecologically sane environment."[22] It urged a "community liberation" in terms of self-sufficiency that would become the "bridge to people's understanding of whole earth ecology."[23]

Counterculture environmentalists in Seattle may not have drawn their ideas directly from PA's "program for self-sufficiency" or the "lifehouse" tips of Ecology Action, but each *Whole Earth Catalog*, paralleling *Reader's Digest* or the *Sears Catalog,* brought together a range of possibilities and offered access to the "tools" with which they would try to balance the urban home with nature. At the time, PA's ideas seemed impractical and fantastical, yet many of them would be incorporated into designs for sustainable cities in the following decades. First, however, they would have to be tested—as they were in Seattle. Aliesan quickly became part of the vanguard of Seattle's own developing counterculture environmental politics, and she and her friend Randy (who managed the Puget Consumer Co-op in the early 1970s) tested new ideas of domesticity in one of Seattle's many communal houses. In the group house that they joined in the Montlake District, they helped incorporate ideas of cooperative consumerism, shared labor, and ecological action into their changing notion of house and home.

The Montlake house had nine members. Despite the "sex, drugs, and rock and roll" image often associated with urban communal living arrangements, the Montlake house was fairly earnest. The housemates shared

SCHOCKEN BOOKS / SB566 / $7.95

BY THE EDITORS OF "RAIN"

RAINBOOK

RESOURCES FOR APPROPRIATE TECHNOLOGY

with Hundreds of Illustrations

Rainbook: Resources for Appropriate Technology (1977) was similar to the *Whole Earth Catalog* but with a more specifically regional focus. The cover image shows the common hope among Ecotopian urbanists to make city living self-sufficient and "green." Tilth Association Records, Cage 602. Courtesy Manuscripts, Archives, and Special Collections, Washington State University, Washington State University Libraries, Pullman, WA.

187

cars and bicycles, they did not own a television, few drank alcohol or took drugs, and of course, they recycled and composted religiously. With the Boeing recession hitting Seattle hard in the early 1970s, the group pooled its resources, and one man did all the shopping and cooking (he would go on to become an accomplished macrobiotic chef). This cooperative atmosphere nurtured a diverse and influential group of counterculture activists who would subsequently develop other projects as part of the evolving alternative economy and life of the Northwest. The "clan father" of the group would later move to Vashon Island to start Island Spring, one of the largest tofu manufacturers in the region. Another member of the house helped to create the popular Bumbershoot arts festival and was one of the first gay men in the country to adopt a child with his partner. In such group houses the politics of environmentally conscious consumption and production were intertwined with broader concerns about loosening rigidly defined gendered notions of labor. And in this context home was far from a closed-off refuge. Such activists sought to erode the wall that separated the private, domestic sphere from the sphere of public political life.[24]

Such communal and group houses served as powerful local testing grounds. They modeled alternative modes of domesticity while explicitly linking consumption to the public realm of environmental politics. One of the more active and influential houses was Prag (People's Revolutionary Action Group) House, on Capitol Hill (another middle-city neighborhood and center of progressive activity), home to Ecotope and Tilth activists, as well as others. Prag House created a more explicitly environmental/ ecological agenda than did many other group houses. It evolved in 1974 around concerns about land use in both Seattle and its hinterlands and the idea of common ownership. The goal of the original house (and the second) was to remove urban land from the market economy, employing strategies along the lines of the land trust concept used in Europe. Eventually, this group expanded beyond its urban communal homes to a rural location, Pragtree farm in northwestern Washington. Each of these original land trust efforts are still operating, and some of the original members are pivotal figures in the cohousing, sustainable agriculture, and organic food brands (such as Cascadian Farm) movement in Washington State. In many ways their work paralleled the earlier work of Walter Hundley and Diana Bower, who created low-cost alternatives to mainstream housing choices as part of the Model Cities program.

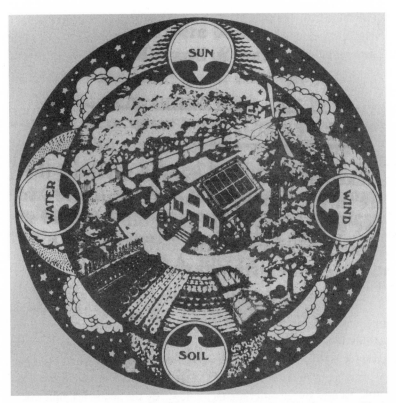

A typical representation of the ideal self-sufficient home found in the pages of *Seriatim* and other counterculture journals in the mid-1970s. From *Tilth Newsletter,* Spring 1977, 29. Tilth Association Records, Cage 602. Courtesy Manuscripts, Archives, and Special Collections, Washington State University, Washington State University Libraries, Pullman, WA.

Prag House and its members were a springboard for another explicitly public effort related to its alternative and appropriate technology efforts. Members of the Prag group, many of whom were also founders of Tilth, developed what they called "Ecotope," an "energy resource center" along the lines of the lifehouse concept described in the *WEC,* later housed at the Home of the Good Shepherd. Ecotope defined its name and mission: "1. A specialized habitat within a larger region. 2. A non-profit research and demonstration corporation working in Seattle, WA since 1974 for demonstration and development of renewable energy and conservation technologies."[25] Through its resource center the group encouraged residents in the Capitol Hill and Central Districts to use solar power, conserve energy, recycle, and redesign buildings to be more energy efficient. They taught the general public to pay attention to the way dwellings and the climate of Seattle and the Northwest could function in concert for increased energy efficiency. The Ecotope group maintained "a specialized resource library featuring conservation and appropriate technology publications and up-to-date technical information on renewable resource technologies";[26] it also had a "community organizer and resource center coordinator" who through workshops and "other community contacts" provided "information on waste recycling, conservation, weatherization, and insulation."[27] Ecotope furthermore published how-to pamphlets, such as "A Solar Greenhouse for the Northwest," which taught Seattleites how to create solar greenhouse extensions to be "used for heat in conjunction with a house."[28] These services brought environmentalism into the home and were designed with the Central District and Stevens neighborhood of Capitol Hill in mind, emphasizing Ecotope's roots in a specific community.[29] After the mid-1970s, however, with the energy crisis in full swing, the long-standing counterculture environmental concerns, especially energy conservation, suddenly had a growing audience of average Seattleites. Ecotope would eventually work with Tilth, the P-patch program, and the concerned gardeners and consumers of PCC to spread the word about greenhouses, composting toilets, and appropriate technology. By the 1990s Ecotope would become a highly successful consulting company; to this day it continues to influence the design of energy-efficient building in the Northwest.

Seattle's local urban policy, too, had begun to favor this sort of work by the late 1970s. By middecade, with the energy crisis settling into the American consciousness, counterculture environmentalist experiments began to seem less utopian and more practical. In the midst of a deepening recession and with drivers enduring long gas lines, the ideas and the related self-sufficiency suddenly grabbed a larger audience, as people began to link their personal habits to public concerns about resources and nature.[30] Seattleites would learn to make this connection by building a sense of self-sufficiency in their homes, but they did so first within the context of neighborhood politics and the legacy of Model Cities. Ultimately, the confluence of counterculture dreams and a rising tide of neighborhood political activism—aided by scarce federal funding—would perpetuate and eventually institutionalize ideas that once seemed radical. In this process, the line between gentrification and liberation could be a fine one. By the late 1970s, when Aliesan applied for her DOE grant, counterculture environmentalists in Seattle, and increasingly, mainstream public and government officials, had already begun to reconceive their city's urban ethos. What began with the preservation fight for the Pike Place Market, the unexpected environmental activism of the Model Cities program, and the neighborhood politics of public space in the mid-1970s became an increasingly mainstream part of the city's plans for revitalization.

"The District That Recycles Itself"

Today the Fremont neighborhood is one of the city's most notorious examples of successful revitalization, or gentrification, an example of a "hip" aesthetic and the "creative economy" in Seattle. More than any other neighborhood in the city, Fremont displayed the triumphs and compromises of counterculture environmentalism with its intense localism, pioneering role in urban recycling, and complicated balance of antipoverty and gentrification agendas. In the late 1970s, though, the area met the People's Architecture ideal of the middle city. Fremont exemplified the merging of counterculture idealism, Model Cities origins, and the eventual institutionalization of a counterculture aesthetic. But it was not always this way in the neighborhood. During the more remote past, in the early 1900s, Fremont was a small logging village northwest of downtown Seattle. One of the farthest stops on the interurban rail line, the village grew into a mostly working-class neighborhood of single-family dwellings situated on the hills around a small town center containing a turn-of-the-century hotel amid two- and three-story brick and stone business buildings. By the 1960s, the old town center, now surrounded by other first-ring suburban neighborhoods (Wallingford, Ballard, and Phinney Ridge) of dense 1920s and 1930s single-family homes, was quite rundown. The working-class housing sat close to the old Gas Works, near railroad tracks, and by the banks of an industrial waterway, giving the area a scruffy mix of light-industrial operations and older architecture. An old three-story bordello—later the Red Door Tavern, before it was moved a block away during the 1990s—was a major landmark in the neighborhood. Despite the combination of quaint and picturesque qualities, Fremont's physical deterioration and concentration of poverty was so extreme that by the early 1970s, when the Model Cities program expanded to the entire city, Fremont became eligible for it. It also became a site of counterculture interventions in community building and the remaking of home.

Seattle's 1970s counterculture was not the first group to seize on the home as a way to restore the city's balance between a changing urban environment and nature. Long before the urban crisis and postwar demographic changes, working- and middle-class residents of Wallingford and Fremont had attempted to negotiate Seattle's emerging modern industrial landscape through the designs of their homes and the associated domestic experience. As did inhabitants of other cities in the late nineteenth and early twentieth centuries, Seattleites made their homes a retreat from a shifting industrial landscape, a protected domestic realm at a distance from the dirt of the new city.[31] Middle-city activists during the 1970s inherited this landscape—the deteriorating first-ring suburbs such as Wallingford and Fremont, where turn-of-the-century Seattleites had attempted to balance urbanism and nature. In houses designed using elements from the Arts and Crafts movement, Seattleites sought a rustic experience in the city. Many of the bungalows that Seattleites restored in the 1970s were structures where this earlier generation had sought to balance the modern urban landscape—pipes and wires, sewer systems and electricity—with the nurturing qualities of home in nature that they were busily cutting out of the nearby forests.[32] Certainly the home Aliesan modeled for her neighbors shared some of these turn-of-the-century Arts and Crafts movement values of "ruggedness, simplicity, naturalness and harmony with its rural wooded site."[33] But she sought a more explicit acknowledgment of the ways in which the neatly divided realms of consumption and production, city and nature, public and private, in fact overlapped. Whereas the earlier version of Seattle living may have reinforced the separation of realms, the new urban homes in Seattle actively questioned this idea. If Seattle's middle- and working-class bungalows of the past had been a retreat, and the postwar reality of home that followed had served to isolate consumers from participation as environmentally aware and liberated citizens, then urban environmentalists after the 1970s would go to the heart of this landscape to reassess its received meanings.

Federal and local government would help. When the Model Cities program expanded in the 1970s, Fremont was the center of a sprawling area designated as the only north-end region included. Model Cities publications from the time marked out an area of urban poverty and blight that encompassed Fremont proper to the North Lake area close to the University of Washington and along the Ship Canal and further north into the Greenwood area. At the time, mostly elderly, white, low-income Seattleites made up the population and focus of the program. The irregular lines that marked the border of the neighborhoods in the program, with Fremont at its center, helped create a kind of origin story for the area and for activists who called the area home. Fremont activists would take their inspiration from the work going on in the Central District and their funding from the same source while they introduced a decidedly countercultural ethos to neighborhood antipoverty programs. In 1972 the city created the North Community Service Center to serve the Fremont area, and through the center the city coordinated federal and local grants for successful neighborhood-building efforts. They rehabilitated housing, provided access to jobs and food for people, and helped organize citizens to ensure "their participation" and "advice" at "all levels of decision and policy" in the program.[34]

In this mid-1970s transitional period, the unintended consequences of federal and municipal government influence, and the neighborhood activism it supported, gained momentum with the newly urgent environmental politics of the time. With an influx of university students and artists, the Fremont neighborhood showed what counterculture environmentalists thought home in the city, and by extension the city itself, could become.

Fremont's evolution also displayed the fine balancing act that activists performed between revitalization and gentrification, public activism and retreat. Fremont citizens took refuge within the boundaries of their little neighborhood and in an identity they derived from recycling—everything from bottles and cans to historic buildings and the family home. The logo of the *Forum*, Fremont's community newspaper during the period, featured the distinctive bright orange drawbridge at the center of the community with the motto "The District That Recycles Itself."[35] Fremont found in recycling a resonant metaphor and identity for a neighborhood (and a city) in the midst of a transformation. Whereas suburbanization had helped diminish the city over the proceeding decades, Fremont's counterculture recycling ethic would show how to consume and thus remake the old landscape. This confluence of causes, along with some interesting personalities, helped create a potent counterculture identity and community model that transcended Fremont's borders. The "district that recycles itself" made waste, conservation, and recycling second nature for Seattleites. Like the people who drove Prag House, Ecotope, Tilth, and the PCC, the jaunty activists in Fremont hoped to call attention to the flow of waste and energy, bringing the domestic realm and the more public realms of community environmentalism into a comprehensive new picture of citizenship.

As did Portland to the south and other American cities in the 1960s and 1970s, Seattle saw a rise in this kind of neighborhood-based grassroots activism. Savvy local politicians tried to accommodate and even co-opt these energies of mobilized citizens.[36] A revitalized small community and alternative press culture that produced several newspapers, including the *Fremont Forum*, the *Facts*, the *Asian Family Affair*, and the *Seattle Sun* ("a weekly newspaper for the Capitol Hill, University District, Eastlake, Montlake, and Cascade Communities," all middle-city locales), gave a voice and definition to these neighborhood concerns, which usually revolved around land use and urban environmental issues. In the aftermath of Model Cities, the antifreeway battles in the late 1960s and early 1970s, and efforts to preserve neighborhood character from the developers and the wrecking ball, Fremont and other middle-city communities dedicated themselves to rebuilding neighborhoods from the ground up. The Fremont neighborhood's community newspaper expressed the emerging sentiment during the 1970s: "It has become increasingly apparent that we are going to have to assume responsibility for our future as a community, both economically and esthetically, and stop waiting for the 'government' or someone else to do it for us. Self-actualization is the key."[37] Echoing the do-it-yourself environmental ethos of Tilth and others, the neighborhood turned inward to a sense of localism.

In 1974, under Nixon's New Federalism, the Model Cities program ceased, ending the associated project in Fremont and reinforcing the sense that neighborhoods were on their own.[38] In the mid-1970s, Richard Nixon proposed his "Better Communities Act," which he explained would give "the lead role back to grassroots governments again." He argued that the time had come to "reject the patronizing notion that Federal planners, peering over the point of a pencil in Washington," could guide citizens' lives better than the citizens themselves could. Instead, he characterized his plan as a "new Declaration of Independence for State and local governments," giving "grassroots governments a new chance to stand on their own feet."[39] It is worth pointing out, however, that in different cities and areas, this policy could have very different effects, especially for African

Americans. Nixon stressed *grassroots government* and not citizens groups. As he did in other environmental plans, Nixon hoped to deflate his opposition, in part by co-opting their rhetoric.

One immediate effect of this new urban policy—signed by President Gerald Ford in 1974 after Nixon's resignation—was to set off a scramble among local governments and citizen-activists to apply for the grants in this new revenue-sharing scheme. In October 1974, the *Seattle Sun* published an extensive guide to Nixon's plans in an article on the Housing and Community Development Act of 1974. In it the mayor of Seattle urged citizens to get involved: "Whether we can solve many of our problems, and whether we can continue to make Seattle a better place to live for all our citizens, will depend . . . on how we use the funds we receive under this new law—and how we use them will depend on you." In an era of diminishing federal dollars for cities, Nixon's New Federalism would make cities fight over smaller pieces of the federal funding pie.[40]

The *Seattle Sun* stressed that things had changed: "It is painfully clear that the biggest and hardest part of planning how to use the block grant will not be deciding what to fund, but what to cut."[41] Development of marginal neighborhoods may have increased following the block grant era, but it also threatened to starve the poorest areas in favor of successful gentrification in other parts of the city.[42] The *Seattle Post-Intelligencer* described this new emphasis as "preventative maintenance in lower middle-class neighborhoods as opposed to the redevelopment of severely blighted areas."[43] In Fremont, however, the residual influence of the federal program fostered a unique entity called the Fremont Public Association (FPA), a nonprofit community organization "providing a variety of social services to meet basic needs of people: a job, a place to live, food and clothing."[44] By 1977 the FPA had grown "from a few people" who had earlier been involved in the now defunct Model Cities program to become a "'a going concern' with a budget of $200,000."[45] Fremont was still a low-rent neighborhood and officially remained in a state of "blight," but its small business and cultural life began to pick up in the 1970s as increasing numbers of artists, bohemians, and counterculture community builders made their ways to this district.

Seattle's middle-city neighborhoods expressed a fierce localism that paralleled that of the emerging antitax and anti-Seattle suburbanites.[46] Fremont residents turned inward, almost barricading themselves in their sense of the neighborhood as a self-determining community. Activists even joked in the *Fremont Forum* that the neighborhood should secede from the city and "open a duty-free port for international trade." "Fremontians" saw the old infrastructure as the future for a distinctive neighborhood within Seattle. The goal, according to the *Forum*, was to "bring Fremont back to its former glory as a self-sufficient, independent, feisty District that saves treasures of the past from the bulldozer and has a soft heart for people in trouble or transients."[47] Downtown and the suburbs were villains to middle-city eyes. In "Downtown vs. Neighborhoods," one Fremont activist, Eli Rashkov, traced the beginnings of 1970s activism. He argued that the revitalized neighborhood movement began in 1968 with the "rapid expansion of community councils, citizen watchdog groups, and 'public citizens,'" as well as the environmental impact statement. Indeed, the organized energies of many neighborhood groups and neighborhood centers evolved along just such a course. Federal and state programs unintentionally activated a public that in turn would change the focus and tone of urban politics in Seattle. And with federal funds declining, according to Rashkov, came a "deepening" conflict with the downtown business district over funding for "eroding services in neighborhoods." In addition, he accused the business district of "detachment from the anatomy of the city" and a "reorientation to the affluent suburbs of Mercer Island, Bellevue and Kirkland," evidenced in part by decisions to build freeways to the suburbs at the expense of the middle-city neighborhoods in their way.[48] Fostering the localism of urban neighborhoods, meeting the desire for self-sufficiency and improvement, required the city's help, and the city government responded, at least rhetorically.

As anger at downtown mounted among increasingly empowered neighborhood groups, successive mayors increasingly offered neighborhood-empowering ideas as mollification. The city's responses indicated the popularity and power of "neighborhoodism" in the period. The city government encouraged neighborhoods to take charge of their environments, going as far as recommending limited traffic flow and cooperative food groups. Wes Uhlman, Seattle's mayor between 1970 and 1978, even employed the counterculture's language and values in some of the city's publications. He made community-building gestures in a HUD-sponsored pamphlet, covered in brown paper to evoke the counterculture, that the Department of Neighborhood Planning distributed to neighborhood groups and public libraries in 1975. "A Sense of Community: Seven Suggestions for Making Your Home and Neighborhood a Better Place to Live" featured tips on crime prevention, house repair, and weatherization; it even encouraged people to ride bicycles. Using the language of the youth culture, the city recommended "getting it together" in the residents' communities by creating "neighborhood co-operatives" for buying food or having "neighborhood sales" to raise money for various efforts. The pamphlet exemplified the unexpected paths that the counterculture environmental ethos of self-sufficiency and do-it-yourself activism traveled during the 1970s. The city even had radical suggestions that echoed People's Architecture in the extent of changes to the urban environment. It encouraged neighborhoods to form "community councils," "land use review boards," and "neighborhood corporations," and it recommended that neighborhood groups plant "street trees" to create "the ultimate living environment" (the pamphlet helpfully recommended appropriate varieties and provided pruning tips). Similarly, the pamphlet gave tips about constructing playgrounds from recycled wood and old tires, and it included extensive diagrams showing how citizens could reshape their neighborhoods by creating traffic diversion islands at intersections, making "block parks," and terminating streets. Such projects, the city argued, would help "instill a feeling of control over the immediate environment," which it conceded was "sometimes lost in the hustle and bustle."[49] In mimicking the counterculture's rhetoric, this pamphlet shows how its ideas had became more acceptable as a strategy of renewal by the 1970s.

As an outgrowth of the Model Cities program, the Fremont neighborhood activists already had a well-developed community agenda that revolved around housing issues and especially blight, or run-down property. American urban policy shifted away from large-scale programs and toward more personal "sweat equity" solutions to urban housing shortages and problems during the 1970s. The federal government's Urban Homestead program (only rhetorically related to Aliesan's DOE grant program) debuted in the mid-1970s, and like Aliesan's homestead it encouraged small-scale and personal solutions while evoking individualist imagery that mixed Old West "pioneering" with New West ideas about reclaiming the middle city. According to one observer of the program, urban homesteading represented "the shift in federal commitments and trends toward stabilization and conservation rather than renewal."[50]

In Fremont, with support from neighborhood activists and local government, Great Society aims had metamorphosed into a more straightforward environmental agenda of recycling and consuming a new kind of urban landscape. An informal poll of Fremont residents in 1977 found run-down property to be their prime concern, above crime and unemployment.[51] Paralleling the situation in the Central District, 83 percent of the residences in Fremont were single-family structures, most of them rentals (64 percent of the neighborhood's 8,600 people rented, whereas 43 percent or residents did in the Seattle area generally).[52] Addressing these figures, the FPA continued the Model Cities agenda and published articles about obtaining low-interest mortgages and tenants' rights and information about shelter for people in crisis. Through the *Fremont Forum*, the FPA, like Ecotope, encouraged more energy efficient houses. Reclaiming middle-city neighborhoods required recycling homes and making them more energy efficient. In successive issues of the *Forum*, editors ran features about the FPA's block grant programs for housing counseling, including the "REHAB program," which was begun by Model Cities activists. The program encouraged everything from "minor home repairs" and loans for repairs and insulation to "full scale housing rehabilitation" and homeownership.[53] Counterculture environmental activists in Fremont, like their counterparts in the Bay Area and elsewhere, had long advocated a decentralized reclamation of the city. By the mid-1970s the federal government and local government, with shrinking budgets and political will, began to mirror the rhetoric that once seemed radical.

As a once blighted community being reborn as distinct, historical, and recycled, Fremont navigated between an emerging gentrification and the older urban crisis poverty agenda. The programs stemming from this effort certainly served the poor as well as the counterculture environmentalists who shared the "pioneer spirit" of middle-city recyclers.[54] The pages of the *Fremont Forum* provided a picture of this changing community, which by the late 1970s had become less poor and more upwardly mobile. With the rents low in the deepening recession, Fremont attracted a burgeoning population of university students and artists in search of cheap housing. Next to news about the Fremont food bank (housed in the old bordello), services for senior citizens, or tenants rights, the *Forum* published profiles of visiting mime troops and artists working in the community. While the *Forum*'s emphasis on fixing up homes in the neighborhood was aligned with the earlier Model Cities antipoverty agenda, it also served the interests of Seattle's counterculture environmentalists and the class of young professionals emerging in the 1980s.[55]

Most of the advertising in the paper reflected the gentrifying side of Fremont, however. Next to articles about home rehabilitation and news about upcoming tenants' rights seminars appeared advertisements for Quiet Man Antiques, which sold "fine eastern oaks"; for the Antique Company, which called itself "a small shop for the discriminating collector"; and for General Antiques, which sold "overstuffed furniture." In addition to selling antiques, local stores, including Stoneway Electric Supply, advertised as cosponsors of the *Fremont Forum*'s housing rehab section; Daly's, a wood-finishing products store, advised Fremonters, "Refinishing wood work and paneling in an older home, or antique furniture you put in it, requires careful, meticulous work. To do it right, please ask us first."[56] With their stock of funky bungalows built between 1890 and 1930, Fremont and the other middle-city neighborhoods appealed to these counterculturalists and do-it-yourself urban pioneers.[57]

Recycling and refinishing the neglected urban fabric expressed a commitment to the ecotopian landscape and matched the new grant agenda of the federal government, blending environmentalist ideas with preservation to justify the early stages of gentrification in Seattle. A reclamation ethic permeated this sensibility. The emerging middle-city aesthetic, and efforts of "Fremontians" of all stripes, signified a shift away from the consumer and environmental values of the suburbs and a recommitment to the city and recycled houses as the human-scaled base of action. By recycling houses in the older city and embracing an aesthetic of the historic, organic, wooden, and funky (Steinbrueck's aesthetic), Fremont citizens demonstrated one aspect of a growing recycling ethos. By the late 1970s, the community even had its own P-patch community garden, a cooperative grocery store, and many of the other accoutrements of the independent, ecotopian middle-city neighborhood. But Fremont had broader ambitions to be an "example to the City and to the entire country" and began to reach out to Seattleites in other neighborhoods.

To take the urban ethos of recycling and ecology to the larger community would require the obsession of one man, Armen Napolean Stepanian. The "Moses" of the curbside recycling movement in Seattle, Stepanian was also Fremont's honorary mayor in the 1970s and a coworker of Frank

Chopp, a founder of the FPA. In the first edition of the *Fremont Forum*, Stepanian, a transplant from New York City by way of San Francisco in the late 1960s, reflected on his rise to power: "To the best of my recollection it was an inauspicious February 27th, back in 1973 when I swamped my 35 esteemed opponents (34 human beings and a black Labrador who was clearly the underdog in the race) to begin my reign of terror as Seattle's only unofficial public official."[58] Stepanian brought a sense of humor to his job as a neighborhood booster, but he took community boosterism as seriously as did any Rotarian. In the *Forum* he listed some of things he had created: the first Fremont craft fair, the Fremont Public Association, and the "Fremont Recycling Station #1" (he also credited himself with "creating the world and resting afterward").[59]

The recycling station was the honorary mayor's center of activity and the base of operations. It was the point of creation from which the ethos of recycling would spread far and wide, eventually helping to encourage Seattle's citizens and their government to establish a source recycling (curbside recycling) program in the city. But more important, Stepanian and his staff of "ecoeducators" and "driver/Ecoeducators" (similar to the environmental aides who trained to work with Pack Rats in the Central District), who manned the recycle trucks, helped to bring home an awareness of the consumer's role in the stream of waste and consumption.[60] Recycling provided Fremont an identity, and Stepanian says that along the way the neighborhood articulated a vision of "household recycling" connecting home in the community to "virgin stands of timber," Northwest rivers dammed to produce energy for aluminum manufacture, and "dependence on foreign nations." In 1977 such sentiments were only beginning to become common. It took a campaign of some effort to convince "each consumer" in the broad Seattle public that he or she was "responsible for the problems of waste disposal" and that each consumer could "contribute to a solution."[61] According to supporters, household recycling, though no panacea, did reveal the connection between nature and consumer decision for the many Seattleites who began learning the activity's private and public rituals.[62]

Beginning in the early 1970s, Fremont hosted an annual arts and crafts fair; Stepanian's and Fremont's recycling identity was born during the 1974 fair. Like other street fairs of the time, the Fremont Fair brought together pottery, juggling, and music. When it began in 1972, there were fifty exhibitors, but by 1977 there were over three hundred booths that sold woodwork, jewelry, leather goods, and sand candles. According to an article in the *Fremont Forum*, recycling started at the fair as "a means of raising a little money to supplement FPA's food and clothing bank." Fairgoers were encouraged to "feed the poor with other people's trash."[63] With the "overwhelming response" at the 1974 fair, Stepanian started Recycling Station no. 1 to serve the Fremont neighborhood. Successive fairs would increasingly incorporate the themes of energy conservation, environmentalism, and waste recycling. By 1977 the popular fair advertised "arts, crafts, tasties, [and] energy saving tips" and filled booths with environmental advocacy groups and representatives of the city who shared the news about energy conservation.[64]

As the PCC, Ecotope, and the FPA encouraged personal solutions to environmental problems, Seattle's concerned citizens and counterculture activists began to pressure the city government to expand services. The city had been searching for better ways to reduce urban waste for some time, making recycling services available to citizens at the south Seattle transfer (waste collection) station since 1970. But few people in Seattle knew about the city's program, and for that matter, few had any notion of source separation recycling before Earth Day and before Fremont made it a crusade. In the years after 1974, while Armen Stepanian popularized recycling at the Fremont Fair, Seattle's engineering department had been struggling to manage overflowing landfills and find new strategies for solid waste disposal. Throughout the early 1970s, various businesses, environmental groups, and concerned citizens lobbied the mayor's office to expand recycling in the city and make it easier.[65] One citizen and recycler, Steve Bender, who had been using the south station since the project began, wrote to encourage the city to advertise the station and keep the area tidy and easy to use. "Many uninformed persons still throw many dollars into the garbage," he said, which could have gone into "building a better city and helping to preserve the Earth."[66]

The interim report stemming from the King County Solid Waste Management Study, performed in 1973, suggested various scenarios for dealing with ever-growing urban wastes. The county officials who dealt with Seattle's waste imagined similar solutions on a grand scale, including massive composting systems that used "sophisticated mechanical digestion" to break down wastes.[67] The authors of the report explored the costs for a citywide recycling system and ideas about how best to shred paper and use magnets to remove ferrous metals. The city measured the market for such recycled products, well aware of the growing crisis. The 1973 waste report concluded, in part, that "for maximum efficiency of a recycle system, there should be home segregation," and that collection of the materials must "concentrate the individual materials to be recycled and bring them to a facility that is most cost effective." Such functions, the report found, were best organized by government but should be "left to the private sector to operate and manage."[68]

With the county recommending private solutions to the problems of solid waste in Seattle, the city turned to local businesspeople, communities, and individuals in exploring the possibility of curbside recycling. Having introduced and popularized the idea in Fremont, Stepanian was a natural choice when it came time for the city to try its first run with recycling. As part of a twelve-month study called the Seattle Recycling Project (SRP), Stepanian and his staff at the Fremont Recycling Station no. 1 sought to perform a subsidiary study that merged, in one gleaming project, all the lessons of urban dwelling that Fremont and the counterculture environmentalists had advocated since the late 1960s. The group received a Washington State Department of Ecology grant to study the viability of curbside recycling in Seattle between July 1, 1976, and June 30, 1977. The usually flamboyant Fremont group conducted a rigorous study and produced a report written in perfect policy language. According to the report, their objectives were to "maximize citizen participation"; "determine the economic feasibility of a solid waste home-based source separation project for the city of Seattle"; "design and implement the operation phase of a model home-source separation project"; and "evaluate all data collected during the project implementation."[69] The group's final report shows how seriously Fremont took their efforts. A handful of other cities in the United States explored bringing recycling into the daily activities of households during the 1970s, but this was the first pilot project in Seattle, and through it Fremont brought its middle-city experiment to a larger public audience.

Over the course of the study, Recycling Station no. 1 sent out letters to participants in the north Seattle test neighborhood, Sacagawea, a residential neighborhood that matched the city's average demographics. Stepanian was quoted as saying, "Just as Sacajawea led Lewis and Clark to the West coast, the Sacagawea neighborhood shows signs of leading Seattle into a new era of the conservation ethic."[70] As did counterculturists, Stepanian drew on stereotypic images of Native Americans as environmentalists. At neighborhood meetings and in news coverage, Stepanian spread the gospel of Fremont recycling: "It is far cheaper to make new out of old than it is to make new out of raw," and "a ton of newspapers save 17 trees."[71] Stepanian and his staff taught the residents of the test neighborhood to place their "ungarbage" of glass, cans, and newspapers in orange bags and place the bags and bundled newspapers on the curbs in front of their homes.[72] They even lobbied Jimmy Carter:

Dear Mr. President,

The first energy saving step that every household should take is recycling at home. Our research project, "Recycling via Source Separation," has demonstrated that with little persuasion, people are ready and willing to sort, prepare, and set out all glass, tin and aluminum containers, and newspaper—the "urban ore" of our society.

Recycling at home lowers disposal and landfill costs, saves energy in remanufacture and shipping, conserves natural resources, and reduces litter. The energy saved by source separation is available immediately, not far into the future. Most important, however, home recycling will prime every American family to respond to other conservation messages.

When recycling begins in the home, consumption becomes a daily reminder to conserve. We strongly urge that home separation recycling be a part of your comprehensive energy conservation program.

Energetically yours,
Armen Napolean Stepanian
Honorary Mayor of Fremont
Director, Seattle Recycling Project[73]

On the scheduled days, Stepanian and his group of paid Fremonters, as well as a group of work-release laborers, would move through the neighborhood collecting the "urban ore" and then take it back to the Fremont station for processing. The final report and news coverage from the time showed a high rate of participation in the program. Residents reported that they supported the program because it was "establishing patterns within the community" and "recycling habits" that should continue; one resident remarked, "It is perhaps a small crumb to throw to the cause of conservation of resources, but also a necessary step both in the minimization of waste and in the higher consciousness of those involved."[74] Fremont helped to raise Seattle's consciousness, tracing the connection between private consumption in the home and the public results of these actions. The city would perform several more similar studies in the years to come before finally implementing a curbside recycling program in the mid-1980s.

Fremont's experience exemplified the mixed origins and evolution of Seattle's counterculture environmental movement. The neighborhood and community-centered activism grew from urban poverty programs, the energy crisis, and the increasing power of neighborhood politics in the city. Ultimately, Fremont's emphasis on its unique middle-city identity would create the platform that the neighborhood would use to transform the habits of Seattle's consumers. These habits and ideas would soon be institutionalized by the city government. The links between domestic consumption and public politics of environmentalisms that Fremont and other counterculture environmental efforts made during the 1970s were further elaborated and amplified in Jody Aliesan's open house.

Urban Homestead

Aliesan's series of open houses offered a counterculture's version of self-sufficiency and environmental consciousness at the domestic level, just as Fremont had modeled similar ideals at the level of neighborhood and city. While the citizens of Fremont popularized middle-city ideals of self-sufficiency and a recycled urban landscape, Jody Aliesan's "urban homestead" synthesized a decade's worth of environmental ideas and gender politics under one roof. With her newspaper articles and in her open houses during the late 1970s, she showcased a rigorous new concept and set of expectations for urban domestic space. In the home, Aliesan and like-minded activists countered the landscape of federally subsidized postwar production—urban renewal, suburban sprawl, and freeways—that they believed had reinforced alienation, gender segregation, and environmental degradation. To replace this landscape, these activists sought to reshape the urban and hinterland environments through consumption and co-operation at the grassroots level in neighborhoods and, ultimately, in the home.[75]

Aliesan wanted to offer ideas that were "inexpensive and accessible to moderate- and low-income people," she said, "commonsense, traditional approaches that worked."[76] For the general public she inspired, however, the open house and Sunday *Seattle Times* columns encouraged radically new ideas of the home yet reinforced more traditional American myths and associations at the same time, just as Fremonters modeled both public work and insular community politics. The idea of domestic self-sufficiency, particularly the idea of the homestead, had deep roots in American history. Aliesan believed the term *homestead* connoted "the place" where "you have taken your stand," where "everybody is working . . . women, men, and children."[77] Aliesan hoped to impress on the Seattle public the often obfuscated relationships among labor in the home, in the city, and in the natural world. But she was also aware of the more negative, isolating, and divisive connotations of the word: "human vs. the wilderness, farmers vs. cowboys, everybody vs. the indigenous people," as she put it.[78] At a time of recession and energy crisis, this evocative mixture of mythic American messages and counterculture ideals may have been especially appealing to the Seattle public in search of change. Aliesan, therefore, had to avoid reproducing the prevailing mythology of the home in following her desire for a domestic space more equal, more connected to the public sphere, and more cognizant of ecology. Aliesan's example revealed the ease with which active counterculture environmentalism could end up reinforcing a quite different ethic of retreat to the garden in the city.

Aliesan and Randy Lee left the group house behind, moving to the north Seattle house that would become the subject of their urban homestead experiment. In 1977, while Stepanian and his ecoeducators were making their rounds picking up recycling not far from her home, Aliesan began inquiring into her own grant work. She decided to apply to the DOE Appropriate Technology Small Grants Program because she felt that people's approaches to the environment were either not changing or not changing quickly enough. She reported a "despair" that she said came "from witnessing how, once again, corporate and government power, greed, and (best case) inertia trumped information, advocacy and alternatives." But, she said, "the 'energy crisis' didn't influence me (except as a possible 'learning moment' opportunity), because I already knew too much about the politics of energy sources and priorities."[79] The decentralized, home-based, and consumer-level solutions that Aliesan and other counterculture environmentalists advocated seemed to match the mood of both the federal and local governments, which by the late 1970s encouraged private solutions to public problems though various programs and grants. Aliesan thought that she could help by using a DOE grant to fund a series of open houses, a sort of neighbor-by-neighbor approach. The grant, which began in 1979, mainly funded the publicity and paperwork necessary to allow her to show the public how she and Randy lived. Based on the popularity of her open houses, the *Seattle Times* invited Aliesan to write a weekly column from 1980 to 1981. For this brief time, in the window opened by the energy crisis, Aliesan had in the grant and the weekend "urban homestead" column two powerful tools for communicating the accumulated lessons of counterculture environmentalism. And readers seemed hungry for her message.

In her house Aliesan created an alternative domestic space in which she laid bare the home as a force of consumption and production shaping the larger environment. Her open houses helped to crystallize these counterculture messages, and she influenced how many Seattleites thought about labor, gender, and consumption and about the ways their everyday domestic activities ultimately influenced the "whole earth ecology." Aliesan drew from a variety of sources for inspiration, including her own childhood working on a farm, her group house, the PCC, and her experiences earlier in the decade. But this work obviously combined the accumulated experiences of Seattle's counterculture environmentalism with the efforts of other activists around the country who had begun to reimagine the home as the center of environmentally balanced urban living.[80] In particular, one powerful vision of self-sufficient living could be found in the Integral Urban House, created by the Farallones Institute of Berkeley, California, as a "research and educational center to develop urban-scale appropriate technology."[81] An integral urban house, according to the institute, combined "principles of biology, food and energy production, and the design of living space and community to create places where one might function without total dependence on an 'artificial,' centralized technology." The challenge, according to these activists, was to "make cities ecologically stable and healthy places to live."[82] Consonant with Seattle's Ecotope and the PCC, the Integral Urban House modeled a vision of the city that charted connections, moving from the domestic realm outward to transform the neighborhood and city. In its 1979 how-to manual the institute published diagrams and plans for their model house, an ambitious ideal of home in a retrofitted turn-of-the-century Berkeley house similar to middle-city houses of the

Pacific Northwest. (The institute even provided before-and-after pictures of the house to show the amount of loving restoration they had performed.) This was no leisure retreat in nature, however; it was an example of self-sufficient, intensive production in the home. The house had an "aquaculture pond" and chicken and rabbit coops for raising food, solar collectors for heating water, and a rooftop garden and large and intensive backyard garden for growing vegetables, as well as compost bins, composting toilets, and honey bee hives.[83] The Integral Urban House, with its lessons about conservation of resources, management of wastes, and consumption (as well as small-scale production), created a set of possibilities for rethinking the role of the house and the urban environment in the larger ecosystem. And though its creators imagined the house as part of the "Integral Urban Neighborhood" and eventually "municipal-scale programs," most of their work concentrated on "self-reliant living in the city."[84]

Aliesan was "aware of the Integral Urban House," which she said she found "impressive" but also "intimidating, foreign, and 'fantasyland' to the average person." She wanted to offer a similar kind of self-sufficiency that would be in reach of "ordinary" people, affordable and low tech.[85] Among Aliesan's various efforts, her energy conservation practices seemed to draw most visitors to her open houses. In the hundreds of postcards that visitors sent, energy consumption and conservation were the most frequently mentioned practical concern. Several respondents reported that since visiting Aliesan's house they had insulated their hot-water heater or were closing their drapes more often.[86] But Aliesan's homestead went beyond encouraging such everyday practical changes by showing Seattleites how their homes interfaced with the surrounding hinterlands. Once in her homestead, each visitor received a promo packet offering quantified data that Aliesan had compiled about her house's energy use and costs, as well as information about her house's impact on the larger environment. Under the rubric "Ways We Conserve Water," she wrote: "Many people believe that because water is plentiful in the Puget Sound there is no reason to conserve it." But, she pointed out, "It takes energy to supply a city with water and to treat and dispose of its wastes."[87] She also detailed her household budget for food, heat, electricity, sewage, water, garbage, and transportation. Her costs were low, she said, allowing Randy and her the chance to live "more flexible" lives and the freedom to take jobs on the basis of interest rather than income.[88] In a section on transportation that compared "efficiency of modes of passenger transport," she noted that she and her partner had stopped driving in 1976, which reduced the "strain on oil reserves, mineral supplies," "street maintenance," and "highway construction."[89] They advocated bicycles. Aliesan's house thus linked cooperative labor and consumption in the home to show how personal freedom was linked to environmental change and how the city could become more balanced with nature. Aliesan's deconstruction of the home revealed flows of energy and waste moving into, through, and out of the home.

In addition to learning about energy and conservation, many visitors were impressed with Aliesan's garden and compost system, a spectacle of self-reliance and food production. Composting food waste seemed to be the most alluring link between personal consumption and household garden production that the open house displayed for visitors. Although relatively few visitors said they would be composting urine, as Jody recommended, many of the over 800 respondents wrote that, for example, they had decided to precompost food scraps. Others announced that they had "built a compost bin" or a "compost pile" and were "composting for [the] first time."[90] In response to the question, "Was the Urban Homestead Open House Useful?" hundreds of respondents described composting and gardening as important parts of the open house's message. Many said they were increasing the production of their gardens, one respondent said he had tripled the size of his garden for the following year, and many had new ideas for solar greenhouses.[91] Seattleites, including visitors to the homestead, were already avid vegetable gardeners in the 1970s, inspired by the PCC and the developing community garden culture in the city. Aliesan's homestead reinforced this development.

Aliesan provided plans, too. Her handouts included the typical plan for her gardening during the year. The garden scheme showed her raised-bed gardens (used in preference to rows), which were "arranged in three mounded areas," with planting schedules from February through September that included everything from bok choy, zucchini, and edible chrysanthemum to Swiss chard and garlic.[92] The plan for this garden was not meant to be an ideal: "It's just what we're doing this year." It was, moreover, organic; she put household waste in the form of compost to work and eschewed toxic substances.[93] Many visitors were intrigued at the sight of Jody's wastes recycled and reused in the garden for productive purposes, a lesson about waste and energy taught at the level of the house.

208

In addition to offering her series of open houses, Aliesan questioned prevailing domestic practices and assumptions in her articles for the *Seattle Times*. From one weekend to the next, Aliesan's "urban homestead" columns appeared among a variety of traditional home and garden columns in the Sunday newspaper. Aliesan's columns explaining her message of organic gardening and reduced energy consumption appeared alongside pieces by the popular Seattle garden expert Ed Hume, who advised older methods, such as using "extreme care" when applying chemicals to remove moss from a roof because they "will eat away metal."[94] Hume's and Aliesan's often diametrically opposed columns appeared beside a variety of other lifestyle articles and home and garden ideas. Aliesan says she was fully aware of the flux of the era, which allowed her voice and vision to be one of several, "contradictions" that "were allowed to exist in relative experimental openness."[95]

Aliesan questioned prevailing assumptions about labor within the home and the gendered scripts they implied by linking changing social relations in the house to a heightened awareness of the effects of urban domestic consumption on the whole earth. As a feminist, Aliesan questioned received ideas about the home as a "'domestic' and therefore 'feminine'" realm.[96] She saw the homestead as a space that "comprises everybody's free development." With the home as the point of departure for her activity, Aliesan attempted to navigate these different ideas of domesticity. In fact, several letters that she received from readers and homestead visitors were addressed to the "Urban Homemaker," and even the DOE played up the idea that she was a "housewife" in its promotions of her project, ignoring her work as a poet, writer, and activist.[97] Her popular columns, however radical, therefore fit nicely within the weekend home section of the *Seattle Times,* which evinced this variety of choices about what home could mean in 1970s Seattle.

One reader who identified herself as the "topic chairman" for the American Association of University Women wrote a letter requesting that her entire group be allowed to visit Aliesan's next urban homestead; it would be opportune, she said, because, the group was currently discussing strategies for resource management and families facing change. She noted, "Of course, the two topics are interrelated."[98] Visitors certainly made the connections that Aliesan hoped to model with her labor. In her columns and open houses, Aliesan made it clear that domestic labor, gender, and ecology were not separate. She performed every function in the house—from making tofu to chopping wood—with Randy as an equal partner (though he never appeared in photographs and receded into the background). Next to Aliesan's articles that showed her chopping wood, wrapping her water heater, and collecting honey on her rooftop in a beekeeper's outfit, the *Seattle Times* published columns with titles such as *Liberated Male* and *Equal Marriage*.[99] In an article entitled "Macho Is Murder," the columnist Jim Sanderson made the connection between high rates of stress in males and early death. He recommended that men "ease up" and that they "did not have to pull the whole load anymore."[100] The *Equal Marriage* weekly column also questioned traditional ideas about the domestic sphere, labor, and relationships. The weekend section, therefore, with its blend of traditional and newer ideas of home, provided a broadening set of choices about what home could be in the early 1980s. Within this variety of traditional and more progressive choices, Aliesan's home put the politics of counterculture environmentalism into the mix of the reimagined house of the late 1970s and early 1980s.

Still, although Aliesan politicized the domestic space, hoping to take the lessons of Prag House and Fremont to the greater public, her message of self-reliance may have inadvertently reinforced existing ideas about the home as an isolated realm apart from the public sphere. And by the 1980s, with both physical and political public space shrinking, such choices were significant. Aliesan offered a critique of prevailing American ideas about the home as a private realm cut off from responsibility, but she did so within an existing political and economic context that increasingly placed atomized houses and personal actions as the remedy to public problems. A reader who identified himself as a "professional environmentalist" voiced this opinion. He reacted to Aliesan's spectacle, calling it a "stereotypical image of wood-burning scavengers wolfing down grapenuts seeking cosmic harmony with Mother Earth." The respondent questioned the "compatibility of the Urban Homestead concept with long-range energy goals and environmental quality." He urged the *Seattle Times* to use its resources to create a more "active and analytical evaluation of environmental and energy issues" so as to inform the public and bring about more "comprehensive and intelligent energy and environmental programs." He questioned her individualized approach and, regarding her project, asked, "Who the hell cares?"[101] Another respondent, a visitor to her open house, wrote, "The Open House reflected a certain lifestyle rather than sensible energy conservation, especially for urban dwellers. . . . I suggest you scrap the tour, call yourself the New Age Puritans, and hold church services instead."[102] Most written reactions to her work were overwhelmingly positive, but the criticism underlines the counterculture's personal and lifestyle-choice political messages that could be at odds with "comprehensive" action and the mainstream environmental movement.

Aliesan's open house and newspaper columns captured a moment of transition from the eruptive political and cultural events of the early 1970s and the liminal space that it opened to the increasingly reactive 1980s, as people in American cities began to reimagine the meaning of the domestic realm in a changing social setting. But just as the counterculture environmentalists had felt ambivalent about the idea of urbanism itself, the idea of home that they created stood as a still shaky foundation for a new social and environmental space. The new Seattle home, with its recycle bins, solar greenhouses, and recycled furniture, seemed more connected to its environment and to nature through acts of self-reliance and the acknowledgment of new consumption patterns, yet the messages of self-sufficiency and survival may have unwittingly served as a blueprint for the retreat from the public realm, be it urban or national, prevalent by the 1980s. The politics of consumer-based counterculture environmentalism had spread far and wide during the 1970s, helping people to trace connections between their private consumption and the public realm of energy conservation and the natural world. The more radical messages pushed hard against the shifting political winds of the 1980s.

"The Changing Western Home"

Among the most bourgeois of bourgeois publications, *Sunset Magazine* ("the magazine of western living") charted the evolution and rising acceptance of the ecotopian house and city by the end of the decade. Each month *Sunset* kept its audience abreast of West Coast counterculture environmentalists' ideas and much of their style. But the magazine also showed the built-in contradictions of their embrace of home. Long an advocate for a regional design and "lifestyle," *Sunset* reinforced and disseminated environmentally conscious home and garden trends through its coverage of the northern reaches of ecotopia. Throughout the late 1970s it published a feature called "The Changing Western Home," a section that claimed to capture the zeitgeist of western living and its latest domestic incarnations. By 1977 *Sunset* regularly published feature articles that encouraged acting on many of the ideas that Aliesan, Ecotope, and the Fremont counterculture environmentalists had advocated, including "solar integrated systems" and other "simple techniques to meet human needs."[103] This was another measure of the counterculture's success and importance by this time—and its failures, too, for *Sunset* encouraged individual solutions with little reference to the concerns of social justice or poverty that preoccupied earlier activists or Jody Aliesan. A far cry from its glossy coverage in the mid-1960s, which featured ideal ranch houses and ideal housewives in ideal kitchens, the taste-making magazine began to resemble the *WEC* with its literature reviews and house improvement ideas.[104] The magazine even recommended that readers consult the Tilth-affiliated *Rain Newsletter*, which *Sunset* described as "possibly the most comprehensive and professionally prepared source book since the legendary *Whole Earth Catalog*."[105] By the late 1970s, the do-it-yourself ideas of the counterculture had made their way to the mainstream of western living, despite the despair of activists such as Aliesan.

In addition to stumping for appropriate technology, *Sunset* encouraged other ecotopian concepts that departed significantly from the tidy images found in its pages only ten years earlier. One article even encouraged readers to plant flowers in the cracks of their driveways: "you can repave, fill the cracks with asphalt patch, or do what landscape designer John Caltin did—he stuffed the cracks with flowers."[106] In similarly ecotopian fashion, the magazine recommended that readers use gray water in the garden, build solar greenhouses, and use salvaged materials—such as railroad ties and "scrounged wood"—to improve the look and authenticity of their houses and yards.[107] These features demonstrate the counterculture's triumph—at least aesthetically. *Sunset* celebrated the recycled structures and scavenged materials of the middle city, as well as the self-sufficiency ethos of the urban home.

But as the stories of Fremont and Aliesan show, each in its own way, another aspect of this self-reliance and self-sufficiency portended an emerging landscape of retreat. Along with espousing a remade domestic sphere rendered more efficient and in harmony with nature, Seattleites and the editors of *Sunset Magazine* cultivated domestic spaces that were increasingly private and protected from the outside world. Just as Aliesan and the Integral Urban House advocated the remedy of self-reliance as a means of survival in the 1970s, the pages of *Sunset* reiterated this domestic landscape but emphasized "privacy from the street." In a discussion of one architect-remodeled house in Seattle, "a faded but well-built Seattle bungalow," the magazine described the "subtle relationship of retreat between the house and its developing evergreen screen."[108] Such articles on landscaping and architecture suggested merging the home with nature in ways that made domestic space more efficient and comfortable but also more cut off from the outside world. Each issue provided detailed diagrams of the gardens it featured, and often the editors suggested ways to create "privacy" for "outdoor living." The pages of *Sunset* displayed an increasing emphasis on transforming the older middle-city yard and home from its previous functions to a more private and intimate space, even while employing "scavenged bricks" and native plants. One article featured a home in Seattle where a driveway was transformed into "a private garden and play space." The homeowners built the new space, which included a Douglas fir log sculpture and pedestal tables made from "old timber segments bolted together," because they said they "couldn't relax outside without being in full view of neighbors and passers-by." Such designs encouraged "privacy from close-in neighbors."[109]

Sunset's articles about environmentally sound building practices and the magazine's celebration of the counterculture's aesthetic of recycled materials and native plants signaled at least a triumph of countercultural style and at best a transformed idea of home in the city—a balance of home, city, and hinterland. But alongside this mainstream acceptance of the middle-city environmental ethos at the close of the 1970s, there germinated the seeds of the 1980s tendency to turn away from public life in cities. The community-oriented and publicly engaged counterculture environmentalism that Aliesan, Fremont, and Ecotope advocated would survive the 1970s, but in popular consciousness and actions it would take a more personal, introverted, and less civic consumer-oriented form. As "morning in America" dawned and a new administration took office, Aliesan stopped writing her column for the *Seattle Times,* and the crisis about energy and scarcity subsided seemingly overnight—mentally, at any rate.

At the beginning of the Reagan era, the filtered sunlight of ecotopia shone down on a new landscape in which backyards bloomed and bags of recycled goods waited on the curbs in front of Seattle homes. But Seattleites had already begun a subtle form of retreat from a landscape of collective activism to an increasingly private backyard-garden realm of personal solutions. This is a story not of black and white, of total success or total failure of ideals, but rather one as gray as the recycled water that ecotopians sprinkled on their gardens. The down side of this greener urbanism may have been its tendency to reinforce a trend toward a more fractured landscape in a city that would become increasingly out of reach to working- and middle-class families, families for whom Fremont had once been home. The domestic realm, having weathered a turbulent decade, would go on to be the pride, in form and idea, of a growing class of professionals who once again made home a fetish. With very different goals, though perhaps no less utopian, young urban professionals would continue the process of gentrification that started with the best public-minded interests of greening the middle city.

Epilogue

Commons

"Park Here—Whispering Firs and Salmon Runs: A Different Sort of Downtown Space." With this tantalizing headline, John Hinterberger, a popular food columnist for the *Seattle Times*, departed from his normal discussion of Seattle's flourishing food scene in April 1991. For two years, Hinterberger told his readers, he and "a small group of people sitting in a room at City Hall" had been quietly eyeing the Cascade and south Lake Union neighborhoods just north of the downtown core. Hinterberger described sitting with his influential friends—including the Pike Place Market preservation veteran and local architect Fred Bassetti—before an artist's rendering, a "map of the Lake Union/Westlake neighborhoods, with a creek running from the south end of the lake toward the center of Seattle." He quoted Bassetti: "Salmon could spawn there"; "tourists and children could come to watch them swimming upstream, jumping up through small waterfalls." Seattle's nature-friendly readers learned of the forty to sixty blocks of real estate that could be converted into green space, encompassing an area running from the shore of Lake Union to Denny Way and from Terry to Dexter Avenue. "Imagine a salmon run in the middle of Seattle," wrote Hinterberger. "Imagine new groves of evergreens where once upon a time there was nothing but old groves of evergreens. In short, imagine the essences of the Northwest—right here in the major urban defoliation of the Northwest." To realize such a vision, the journalist argued with little

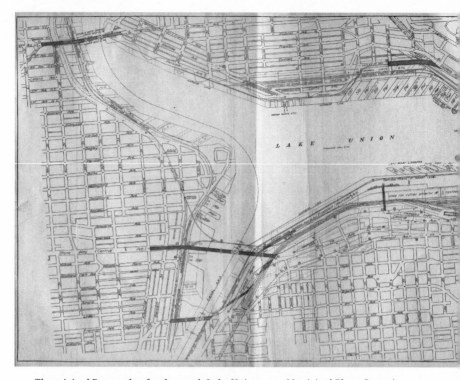

The original Bogue plan for the south Lake Union area, Municipal Plans Commission, 1911. Maps Included in "Plan of Seattle / Report of the Municipal Plans Commission Submitting Report of Virgil G. Bogue," item no. 619. Courtesy Seattle Municipal Archives.

irony, would require doing to Seattle "what Baron Haussmann had done to Paris" in the nineteenth century, and he proposed that the city "bulldoze marginal areas, create boulevards, open up the streets to the sky. AND BUILD A MAJOR PARK."[1]

The marginal areas to be bulldozed included the Cascade neighborhood, which sits just north of downtown on the south shore of Lake Union. By the 1990s few Seattleites knew of Cascade or its history. This area encompasses a section of the city from Denny Street in the south to the shore of the lake at the north and from the eastern slope of Capitol Hill to the area near today's Highway 99 to the west. At the turn of the last century, Cascade was a small mill town, and until recently it contained an assortment of ornate Italianate homes, working-class houses, and a mix of light-industrial and small businesses. The neighborhood still reflects a

216

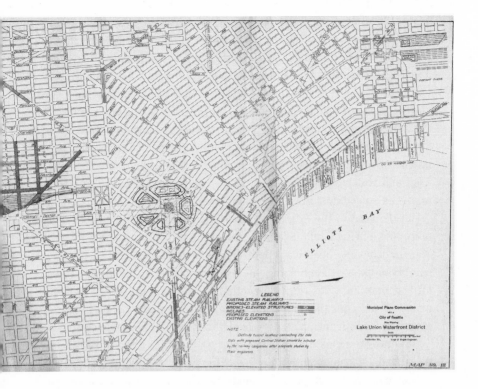

working-class character and the cultural remnants of the Greeks, Swedes, and Russians who once lived close to downtown. The neighborhood had a long history as the focus of failed grand plans as well. In the late nineteenth century R. H. Thomson, a city engineer, and Virgil Bogue, an urban planner, had imagined "a major linkage tying the downtown, Denny Regrade, Seattle Center, South Lake Union and Cascade neighborhoods together in an attractive, complementary and coherent whole." The "conflux" of Seattle, as Bogue had called it, seemed once again within reach in a new, supposedly greener age.[2]

This vision of green space soon took hold among a broad swath of Seattleites. Appealing to a readership with increasingly refined tastes, Hinterberger offered a prize to the person suggesting the best name for the proposed park: a wine and cheese picnic with the columnist. Readers suggested "the Promenade" and "Evergreen Acres," among other names.[3] Hinterberger encouraged readers to write to the mayor and city council with their visions of "running paths," "picnic tables and seasonal gardens," and a "grand boulevard and maybe a speakers corner."[4] As the plan's contours broadened, it became clear that the project would include not just a plan for increased green space but a massive redevelopment scheme for the dormant section of the city. By 1992 the *Seattle Post-Intelligencer* editors agreed that the area was ripe for redevelopment.[5] The redevelopment might solve traffic problems, "better utilize" a neighborhood of "underused and widely dilapidated real estate," and even solve environmental problems caused by storm water runoff that affected Lake Union.[6]

In the flush of the "Microsoft moment," a tech boom that echoed the mid-1960s Boeing boom, Seattle's citizens once again debated a redevelopment scheme of massive proportions. After two-and-a-half decades of neighborhood-based politics and incremental planning, and having transformed from an industrial to a postindustrial city, Seattle now contemplated urban plans that appropriately revolved around a popular mixture of salmon and streams, public space and "sustainable development."

By the late 1980s the previous decade's creative energies had mostly dissipated, pushed to the margins by an increasingly privatized landscape of consumption. In this new context the benighted little neighborhood near the shores of Lake Union, and the contemporary debate over sustainability that it would inspire, suggested the mixed legacy of Seattle's place making between the 1960s and the late 1980s. The debate reflected some of the essential and unresolved conflicts of the urban environmental politics born over the previous two decades. Cascade itself had little in common with the far more historically significant market or Central District. Nonetheless, in the 1990s the neighborhood evoked Seattle's pained narrative of place making and underlined the unresolved contradictions at the core of postwar urban environmentalism and its relation to class and power in the production of urban space.

"Urban Villages," Old and New

Despite the city's economic and cultural recovery, many of its urban problems were as acute as those of any other city in the 1980s. At the end of the Reagan years Seattle had its own big-city homeless crisis, a dominant symbol of urban failure in the period. Many Seattleites worried about crime rates and security, as well as the shortage of affordable housing and corporate development in their own downtown. For example, a series of proposed ordinances debated during the early 1990s showed the Seattle electorate's concern over dealing with panhandlers and cleaning up the image of downtown to help businesses there. Despite their success at creating a green ethos, Seattleites remained nervous about public spaces.[7]

After a loss of over 63,000 people to suburbanization and white flight between 1960 and 1980, the city once again appeared to take its rightful place as the cultural and economic hub of the region—all that 1950s planners had hoped for.[8] This renaissance was due in no small part to the creative efforts of neighborhood activists beginning in the 1960s. Even during recession and the rapid growth in the 1980s, Seattleites felt they had perfected a blend of new service industries, culture and arts, and community—the stereotypical yuppie culture that urban sociologists would later describe as the "the rise of the creative class" and the recipe for urban success stories around the country.[9]

Seattleites and Washingtonians hoped to preserve the gains of the previous decades. To resist sprawl and protect the city's "livability," they began more significant attempts to control growth in urban and hinterland environments. Seattle's nearby neighbor, Portland, passed an urban growth boundary in the late 1970s and by the 1990s began to put teeth and planning authority behind the decision. In 1990 Washington's state legislature passed the Growth Management Act, which asserted that "uncoordinated and unplanned growth, together with a lack of common goals expressing the public's interest in the conservation and the wise use of our lands," posed a "threat to the environment, sustainable economic development, and the health, safety, and high quality of life enjoyed by residents of the state."[10] It was during this time of urban dread and urban triumphs that the next grand plan for the city materialized, this one targeting the southern shores of Lake Union: the Seattle Commons.[11]

"Some heavy hitters are backing a grand plan for the corridor from Westlake Center to Lake Union," the *Seattle Times* announced in February 1992. That year, urban movers and shakers—some well known from the city's past (such as the lawyer Jim Ellis, the organizer of Forward Thrust in the late 1960s) and some later to become even more well known (the developer Paul Schell, who would go on to be mayor during the WTO protest)—put their significant public weight behind the plan to redevelop the dormant "conflux" of the city, forming the Committee for the Seattle Commons to do so. This new committee recommended that the city create the park, though preserving employment in the area while doing so, and create a planned neighborhood on either side of it. Their first plans called for 15,000 units of moderate-, low-, and "market-rate apartments and housing"; a new transportation system in the area (rerouted streets and a rail link); and finally, an effort at "cleaning up Lake Union" with new sewer and holding systems.[12] The Commons committee imagined a fully formed ecotopia near the shores of Lake Union, a miniature version of the best features that Seattle neighborhood activists had perfected—on a shoestring budget—between 1960 and 1980.

Because of its broad scope—at once a transportation plan, an economic stimulus for downtown, and a plan for density—the Commons proposal quickly piqued the common concern of multiple stakeholders. As things moved along quickly, boosters worked hard to avoid the pitfalls of the past and drew—as the market preservationists had—on the expertise of University of Washington faculty and students to help expand and coordinate the plans.

The intellectual underpinnings of the Commons plan could be found in the popular early 1990s planning philosophy of "New Urbanism," a movement that owed a great deal to ideas that community activists and environmental heretics first tested in the 1960s and 1970s. In the spring of 1992, the University of Washington's Department of Urban Planning collaborated with others in a charrette to brainstorm plans for development and to showcase the new "urban village" concept. Douglas Kelbaugh, the leader of the effort and then a professor at the University of Washington, would later publish *Common Place: Toward Neighborhood and Regional Design,* based in part on the experience. The book became a major work in the emerging movement of New Urbanism.[13] Peter Calthorpe, another influential regional planner of the period, a cofounder of the Congress of New Urbanism, and a close colleague of Kelbaugh, shows the clear links between this popular 1990s planning movement and the 1970s counterculture. His 1983 book *Sustainable Communities: A New Design Synthesis for Cities, Suburbs, and Towns,* coauthored with Sim Van der Ryn, an architect who helped create the Integral Urban House in Berkeley, encapsulated many of the once "heretical" ideas that would return at the end of the century, embraced by the Seattle establishment. Calthorpe's *Next Ameri-*

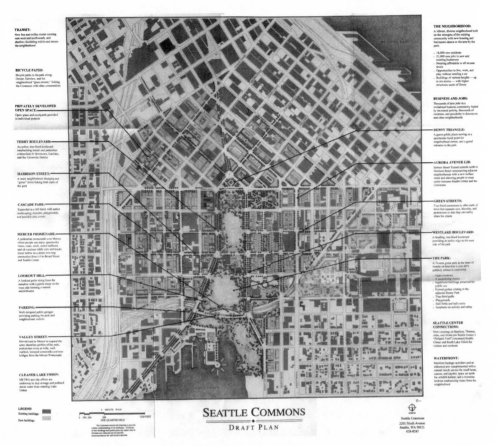

SEATTLE COMMONS
DRAFT PLAN

"The Seattle Commons Draft Plan," covering much of the same area featured in the Bogue plan. The area on the left side of the map is the Cascade neighborhood, 1993. Maps Collection, item no. 338. Courtesy Seattle Municipal Archives.

can Metropolis: Ecology, Community, and the American Dream, a further elaboration of ideas he first tested as a member of the Farallones Institute, appeared in 1993 in the midst of the Commons debate, providing another powerful articulation of the new planning philosophy. By the early 1990s, then, Seattle was a hotbed of New Urbanism, as its proponents watched to see whether citizens there would implement this philosophy on a large scale.[14]

The group of visiting architects, landscape designers, students, and faculty set their sights on the area, proposing different scenarios for development of green space and urban villages in the south Lake Union neigh-borhood.[15] In *Envisioning an Urban Village,* produced for city officials and citizens in 1992, Kelbaugh detailed his own arguments for the urban vil-lage concept, a plan that would use the Commons neighborhood to take the city's fair share of urban growth rather than spill more into suburbs of King County. In addition, he argued that urban villages would be econom-ical, "walkable," "transit-friendly," "neighborly," "exciting," and "sustain-able." The state and the city were at a "historic crossroads," he believed. In order to maintain the region's "livability and scenic beauty" and to "provide an alternative to Los Angeles and Houston," Seattle should cre-ate these urban villages: "We must design and build new models of afford-able, livable and sustainable lifestyles."[16]

From the outset, the "heavy hitters" behind the plan did their best to anticipate and accommodate the concerns of potential critics. Most of the insiders had been part of the earlier struggles over Pioneer Square or the Pike Place Market and were familiar with rabble-rousers. These one-time champions of preservation would have to grapple with the strong an-tidowntown attitudes that neighborhood activists and the counterculture had cultivated over the previous decades.[17] They would also face what lo-cals euphemistically called the "Seattle way of doing things": the slow, consensus-driven process by which Seattleites seemed to make deci-sions—or utterly fail to. "Public involvement," according to the Commons' booster Paul Schell, is "Seattle's greatest weakness, and its strength." "Most American cities," he maintained, are "so far gone nobody cares any-more." In Seattle, however, "everybody cares": "We're close to strangling ourselves with good intentions."[18]

Money became no object in October 1992 when an anonymous bene-factor opened a $20 million line of credit to be used for land acquisition in the south Lake Union area. The money kick started the next intense plan-ning phase, allowing boosters to continue their public relations campaign, hold meetings, and make glossy maps. They did not need to rely on the city's resources. A far cry from the activists who had depended on small federal grants for their urban renewal projects of the past, these elites in-side and outside city hall now had private money and all the best inten-tions—not to mention the greenest.[19] No one would be scrambling for fed-eral money or condemning blighted neighborhoods, as had occurred in the mid-1960s. With the new language of "sustainability"—and having learned the lessons popularized by the grassroots and counterculture activists of the 1960s and 1970s—they hoped to create a model of public process and public space in the heart of the green city.

New Urbanists, then, saw the crumbling neighborhoods of south Lake Union as a staging ground for their sustainable urban villages, but during the seed time of this movement, in the 1970s, students, activists, and artists had seen the Cascade neighborhood in a similar though different way. Twenty years earlier activists had tried to make their case for a version of the urban village, but with dwindling public funds and in the far more cynical political atmosphere of the Nixon era. When the city had sought to rezone Cascade, the neighborhood came to play a minor but auspicious role in the drama of mid-1970s urban environmental politics. This earlier era left a powerful imprint, a memory of resistance that would come back to haunt the Commons boosters.

The underground press described an embattled Cascade as early as 1973. Activists interviewed for a *Northwest Passage* article in 1973 feared that if the planners "had their way, . . . they would pave over the Cascade neighborhood." The Bellingham-based alternative weekly described a newly founded "Community Council" and the neighborhood's "struggle for survival." The paper characterized the community as being "still very much like a small town": "In many ways, most people know each other, share common interests, and are all friends with each other and neighbors in the true sense of the word, something you don't often find in cities nowadays."[20] "We need people who can talk to the elderly, who can organize food-buying clubs, and who can become familiar with rezoning laws," said one activist. Such an effort required serious people to "put in some slow, hard, unpretentious, and non-rhetorical organizing," helping to create the "people's control of their own community."[21] A year later, the *Seattle Sun* similarly chronicled Cascade's woes, reporting, "Cascade's remaining population of about 900 souls clings to their steadily decaying community like shipwrecked voyagers clinging to a storm-swept atoll."[22] "In the past five years" the *Sun* explained, "a band of urban activists . . . drifted into the community, attracted by low rents for housing and the opportunity to experiment with alternative lifestyles." This group "formed the nucleus of what is now the Cascade Community Council [also called the Cascade Neighborhood Council] and set out to challenge the previously unchallenged assumption that Cascade would 'go industrial.'"[23] Cascade's council also produced its own "ten-year plan" that, evoking both Saul Alinsky and Stewart Brand's *Whole Earth Catalog*, stressed "shelter," "identity," "cooperatives," and "struggle." Council members hoped such changes would unify the neighborhood and help make it "part of the broader people's movement throughout the society . . . aware of the interrelated wholeness of society."[24] Long before the Commons debate, Cascade's defenders had developed their own version of the urban village.

A young activist named Frank Chopp epitomized the broadening neighborhood and early sustainability movement in Seattle during these harder times. The son of a Croatian shipyard worker in Bremerton, Chopp graduated from the University of Washington in 1975 and soon thereafter moved to Cascade to protest injustice downtown in the Saul Alinsky mode.[25] Chopp said, "I saw a lot of poor people being displaced by development."[26] An outsider to the neighborhood, he soon settled in to help demonstrate the problem of housing downtown. Chopp rented a parking space in a lot next to the Immanuel Lutheran Church and built a geodesic dome there (a very countercultural gesture)—with electricity provided via an extension cord from the church basement. Chopp would later become a crucial leader in the Fremont neighborhood—describing himself as a "wild person from Fremont"—where he worked, alongside Armen Stepanian, for the Model Cities program. Drawing on his work as a community organizer, he helped found the Fremont Public Association, a group that assisted low-income, elderly, and homeless people in the community.[27] Chopp shaped a career in politics that would take him to the state's House of Representatives. More important, he mentored the next generation of housing activists, including those who would work to defend Cascade during the 1990s.

In search of wholeness and small-scale community, Cascade activists in the 1970s worked within a difficult moment of transition in urban policy and funding. Following the previous decade's so-called urban crisis, then, community pioneers in Cascade who sought to reclaim urban space in the 1970s did so within a new framework that favored private efforts over public. The Housing and Community Development Act prohibited use of block grants for social services and instead emphasized the restoration, preservation, and conservation of urban environments in any part of the city. Nixon's policies importantly shaped the gentrification politics in Cascade and elsewhere. Most important perhaps for Seattleites was the act's explicit support of urban environmentalist activities such as "the preservation or restoration of historic sites, the beautification of urban land, the conservation of open spaces, natural resources, and scenic areas." The act emphasized these new aesthetic goals over antipoverty efforts or affordable housing. Nixon shaped the act to please conservatives who favored limited government and decentralization, yet the funding structure also benefited gentrification in places such as Seattle, where people had the capital and political power to reshape their urban environment. In Cascade, as in Fremont at the same time, gentrification was aided by such urban policies.

By the early 1980s Seattle's newspapers and weeklies reported favorably on the new "real estate rush" and a "Seattle in transition."[28] In "Neighborhoods Come Alive," the *Seattle Post-Intelligencer* linked this upsurge in real estate values directly to the "democratization of city politics" in places such as Cascade, where community organizing was relatively potent: "Today, City Council hearings on subjects ranging from shoreline planning to housing investment policies are packed with representatives of community councils, neighborhood land-use groups and all manner of organizations."[29]

In 1977, however, the nucleus of youth seeking alternative lifestyles in Cascade had used their expertise as organizers and committed city dwellers to commission the rezoning appeal to the city. According to the 1977 report, these activists were "willing to put up with small units, deteriorating buildings, and a noisy, mixed-use area for the advantage of paying low rents and being close to downtown and their place of work." Although the activists hoped the area would not "go industrial," they certainly cherished the edgy mix of residential and industrial uses, and they liked Cascade's proximity to downtown Seattle: "The area is well served by public transportation and is extremely convenient to downtown and other major destinations within the City," according to the report. For these activists Cascade epitomized the possibility of neighborhood living downtown: a model for mixed-use urban living, a place where work and affordable housing might coexist in the city.[30]

Dueling Sustainability

The Seattle City Council formally endorsed a retooled Commons plan at the end of 1993. In the following years, the debate over the park and the details of the plan responded directly to a variety of public criticisms hashed out in public meetings. But it became clear during 1993 that much of the opposition—though not all—revolved around the future, history, and invented traditions of the Cascade neighborhood.

One of Frank Chopp's protégés, John Fox, of the Seattle Displacement Coalition—a group that fights for the housing rights of the homeless and working poor—helped lead the vociferous opposition. Fox amplified the plight of the poor in the neighborhood for a citywide audience. In doing so, he effectively held the feet of sustainability advocates to the fire. After a public meeting in 1993, Fox called the plan "a giant game of Monopoly" and charged, "The corporate community and the Commons . . . are telling this community what's good for them, rather than working with them first."[31] In Cascade during the 1990s activists would create a counter sustainability that contrasted sharply with the sweeping development plans for the neighborhood.[32] The longtime Seattle activist and writer Walt Crowley concurred with these concerns at the time, reminding the public that they should not watch as "long-established urban priorities" were "shouldered aside to make room for a plan that was not even a twinkle in anyone's eye" a few years earlier.[33] Other Cascade-based groups soon organized to challenge the plan in 1993, including the Cascade Residents Action Group (CRAG), one of whose members, an eighty-nine-year-old resident, explained, "The truth of the matter is we've got a good community here and we want to save it."[34] The effort to block the Commons also revived the dormant Cascade Neighborhood Council, which had taken shape during the 1970s.[35]

On the eve of the 1995 vote, the Cascade Community Council—the leading opposition—collaborated with yet another group of graduate students and faculty at the University of Washington, this one calling themselves the "Center for Sustainable Communities," to produce a document stating an alternative version of sustainability, "The Cascade Neighborhood Sustainable Community Profile—1995," which was a "small-is-beautiful" doppelgänger of the Commons plan. "The Commons and Cascade both pay homage to the *urban village* and the sustainable community, but in very different ways and with radically different approaches."[36] In the early 1990s, the report stated, "Cascadians were suddenly awakened by all kinds of people showing interest in their once forgotten part of town. While speculators started looking at Cascade for opportunities, activists of all kinds came to the community to get involved; some to fight the Commons, others to protect affordable housing, others to explore the potential for a home grown Cascade version of the sustainable community."[37]

The neighborhood crafted a well-formed picture of urban ecological and neighborhood preservation. Such ideas had percolated since the 1970s, but the vision for Cascade was perhaps the most clearly articulated and politically informed conception of these connections for an urban neighborhood in Seattle—a synthesis of urban landscape and environmentalism.[38] Supporters of this alternative sustainability expressed "a concern for the ecology, but not in the traditional sense of environmentalism." As had their counterparts during the market debate, they suggested "conserving resources, and becoming more efficient, but more importantly . . . building and strengthening the local community, both neighborhood systems and identity."[39]

In 1995 activists in Cascade began their plans to build the neighborhood's own miniature commons—a P-patch community garden—in part as a protest against what they viewed as the extravagant ideas in the Commons plan. Their garden offered a sign of permanence (or invented permanence) and community rootedness. Just as Hinterberger and other Commons supporters coaxed voters with visions of salmon and groves of evergreens, Cascade activists returned to Seattle's quintessential symbol of community for its ideas of sustainability. They reported, "Cascade is fortunate in that the Seattle Parks Department is already taking steps to acquire a vacant lot in the Cascade Playground block for use as a future P-patch."[40] The history of 1970s place making returned in the Cascade neighborhood to challenge, in its humble, yeoman farmer image, the best-laid plans of the Commons planners.

As the 1995 election drew near, an investigative news story revealed that Paul Allen, a cofounder of Microsoft and one of the richest men in the world at that time, was the anonymous donor who had provided the $20 million line of credit for the Commons project. His involvement, along with that of other high-tech innovators and millionaires, including Craig McCaw, confirmed for many the elite flavor of the plans, contributing to the sense that this new "creative class" was guiding the future of Seattle's urban space.

It took two elections to kill the Commons. In 1995 voters rejected the first funding plan and design. The city redrew the plans and reduced the price tag, but voters again turned down the plan. Despite the incredible

A Cascade neighborhood street scene showing the mix of light industry, worker's houses, and Russian Orthodox Church, Nov. 2, 1953. Planning Photographs, Miscellaneous Streets (Yale and Fourth), item no. 44776. Courtesy Seattle Municipal Archives.

momentum and organizational expertise, as well as planners' genuine efforts to organize public opinion from the start, the plan failed decisively at the polls. Days before the second and final vote on the Commons, the longtime Seattle watcher David Brewster captured the divide: "Viewed in one light, then, the Commons seemed like the perfect crusade to revive and push forward the city-building spirit that has been hibernating since the saving of the Pike Place Market and Pioneer Square in the 1970s. Viewed another way, however, the vast army of allies (including developers, society types, and academic empire builders) seemed too good to be true. Without a real debate—and the early ravishment of the dailies didn't help—the taxpayers and the unbeautiful people began to suspect a railroading, a heavily lawyered plot."[41]

228

Although the opposition to the plan was widespread in Seattle, resistance to the Commons plan found a human and rhetorical epicenter in the Cascade neighborhood. The contested idea of sustainability sat at the center of the Commons debate as Seattle grappled with the tangled interrelationships of urban ecology, economic prosperity, and democratic process. Cascade activists saw urban destruction where old preservationists saw a "city building spirit." As one observer put it, "The City of Seattle should be building upon the existing South Lake Union neighborhood, celebrating its concentration of historic buildings and locally owned businesses, rather than seeking its destruction. Interesting cities are created over time, not bulldozed into importance. Under no circumstances should this rich mix of our city's life and history be destroyed."[42]

Rather than reaffirm the activism of the 1960s and 1970s, however, the success seemed to signal its final eclipse. The 1970s had altered the spirit, shape, and process of urban planning, and by 1990 the change seemed permanent. Seattle voters rejected a plan hatched by elites, private capital, and a city's righteous intentions to limit sprawl and create a livable city. The voters rejected the plan even though it reflected the collected wisdom of urban planning and process since the 1970s, including the green sheen of sustainability. Voters chose the more intimate, incremental, small-scale, and organic vision of place instead of an ambitious public space and urban village at the heart of their city.

In the days after the special election, a *Seattle Times* headline read, "With Defeat, All Eyes on Allen—What Will He Do with His Commons Area Property?" After the plan failed, Paul Allen was left owning much of the real estate around the south Lake Union neighborhood. Had voters passed the second Commons ballot measure, he would have given the land to the city as a gift. His holdings in 1996 amounted to 11.6 acres of prime urban real estate. With voters turning down the $167 million plan for urban villages around a forty-two-acre park, the paper reported, "The city is backing off for a while."[43] When planners and city council members had imagined the urban village, it had centered on a park but also on an idea of the next gold rush, this one in biotechnology. As early as 1990 firms in the biotech industry—including the Fred Hutchison Cancer Research Center and ZymoGenetics—had developed their campuses along the shorelines and in the neighborhoods of the south Lake Union area.

Without a large park to make the area cohere as a public space, and without the eventual infusion of public investment, the vision of a biotech boomtown seemed in question. Exhausted by the public process, many elected officials, volunteers, and supporters receded in disgust; the "Seattle way of doing things" had taken its toll. In any event, when the dust settled, it was clear that the fourth-richest man in the world owned a substantial portion of downtown's future. In one sense, the worst fears of activists and other opponents of the privately funded massive public space had come to pass.

In the ever more privatized atmosphere that the election indicated, however, the future of the Cascade neighborhood seemed ironically more ensured. The little neighborhood had formed the center of resistance to the big plans of the city and elite planners. To the chagrin of other neighborhoods and neighborhood leaders, a great deal of money, expertise, and attention had been lavished on the area during the Commons debate.[44] The *Seattle Times* reported in late May that the city would designate Cascade "a historical neighborhood," with all the protections that implied: "The Cascade neighborhood . . . is likely to get new development guidelines to allow more apartment buildings with roof-top gardens, more mom and pop stores and higher office buildings. And city planners probably will take a fresh look at the overall vision for the 560 acre area."[45] After all the planning on both sides, all the neighborhood studies and neighborhood histories inspired during the Commons debate, all the invented traditions and real ones, Cascade emerged as a distinctive place. With its character and defending voices, it was now assured of being a central feature of any plans for the future of south Lake Union. Its very funkiness and marginal status had become its means of survival and source of allure to the next generation of urban pioneers. The people of Cascade would still have to fight to make sure the momentum for fair and affordable housing and preservation continued, but in 1996, they were in a strong position. The neighborhood had been recognized by the city as a peculiar cultural asset, not a blank slate.

By the following spring, Allen had begun to reveal his intentions for the area. He would continue with the "urban village" concept and become a champion of the city's gamble on a biotech future. The *Seattle Times* called it the "dawn of Seattle Commons III—only with laboratories instead of fountains." Allen moved forward. But now it was a wholly private venture. Allen's developments in the area would require the same land-use review that any other developer would face, but gone was the leverage and heightened oversight that the public aspect of the project had once created. The transformations afoot in south Lake Union were no longer an overt private/public partnership. In the spring of 1997 Allen's company began renegotiating lease agreements and raising rents on his properties in the area, indicating big changes to come.[46]

By 2003 Allen had acquired even more land in the area. The *Seattle Post-Intelligencer* reported that Allen and his Vulcan investment firm possessed at least forty-five acres. Allen's redevelopment of the south Lake Union area did not follow the democratic and decentralized mode of the earlier efforts at grassroots urban renewal, yet his development there would nonetheless show the distinct influence of the previous thirty years of Seattle's striving toward sustainability. Seattleites had in the past attempted to create utopian spaces, the 1970s versions of urban villages; Allen would now take this method of planning and place making, pioneered in post–Earth Day environmental movement, and use it to sell his remade landscapes in south Lake Union. Cascade and south Lake Union became Allen's test case, the neighborhood where all Seattle's experiments with sustainability and post-1960s urbanism might come together.

In an unlikely way, then, Allen is the ultimate do-it-yourself regionalist, the apotheosis of one version of the counterculture's values of freedom and futurism popularized in the *Whole Earth Catalog* or *Ecotopia*. Allen built his fortune on a vision that is distinctly a product of the post-1960s age of personal and appropriate technologies, self-expression, and the dream of decentralized digital futures.[47] Allen is also intensely local, focusing much of his business and philanthropic energy in the Pacific Northwest. One of his more famous Seattle efforts is the Experience Music project. Originally imagined as a "shrine to Jimi Hendrix," the locally born guitarist,[48] the project, on the grounds of the Seattle Center, has grown into a museum celebrating the region's music culture past and present.[49]

A salient example of Allen's repackaging of counterculture urban sensibilities was on display by 2004 in the Alcyone Apartments. Across the street from Cascade's young community garden, Allen's company, Vulcan, and Harbor Properties demolished several old wood-frame homes that were not designated historically important to build a large apartment complex. The Alcyone defined the west side of Cascade. After 2000, similar apartment houses, as well as low-income housing units, were built to serve the evolving biotech village and to satisfy housing advocates and the city's attempts to include low-income housing with new construction.

The advertising literature and Web site for the Alcyone Apartments provide interesting evidence about Seattle's evolving urban nature aesthetic in its fully privatized twenty-first-century version. A glossy folder of pamphlets offers a narrative about an urban lifestyle that seems as complete and contained as that of any retirement community in Florida. The pamphlet even includes make-believe journal entries and a calendar of events that the potential resident might experience in one week. Sunday: "touch football in the park"; Monday: "Spanish class @ People Center"; Saturday: "Kayaking L. Union." The fictional journal entries were designed to appeal to young twenty-something urban professionals, prompting them to leave the commute behind and move downtown: "My home is now a factor in determining what I can do instead of what I can't. Some days I pull my kayak down to Lake Union after work and other evenings I take in a show at the Moore Theatre. I'm five minutes away from water, downtown and even a climbing wall that rivals Smith Rock."[50]

Images next to the text show vigorous young people working on laptops, walking dogs, playing football, and gardening—one pamphlet even has tips for growing tomatoes. Promotional photos on the Alcyone Web site show a view of the building through a row of blooming flowers in the community P-patch across the street. The urban garden ideal, the notion of health and nearby nature, and organic community—the same hopes of the counterculture and neighborhood activists in the past—sat at the center of the plan to market Cascade and entice people to live downtown. And if the P-patch across the way from the Alcyone was too public for some, the building had its own private version in the building's "roof-top Pea Patch garden with recycled rainwater irrigation," which provides "green space, reduces potable water consumption, and mitigates storm water runoff from the roof."[51]

The heretical-sounding goals of in-city agriculture, recycled building materials, and water conservation—counterculture environmental goals of the likes of Ecotope, Tilth, and Jody Aliesan—occupied the center of Seattle's newest urban vision for living. Indeed, by the end of the century companies such as Ecotope and other Tilth-associated businesses had blossomed into larger, more mainstream concerns, consulting with builders interested in incorporating such technologies. The "sustainability fact sheet" and other pamphlets provided to potential residents describe in detail how Alcyone is "a healthy place to live," where the units have "energy saving windows and doors" and "recycled water is used for watering the rooftop Pea-patch." The entire building is sustainable and LEED (Leadership in Energy and Environmental Design) certified: "the first LEED certified residential building in the Cascade Neighborhood, and one of the first in Seattle."[52] Allen's prototype for the planned urban development used this sense of sustainability as its chief draw, combined with the feeling of an "authentic community."

Nonetheless, Allen's attempt to provide the latest in environmentally sound developments to rival the dreams of earlier counterculture environmentalists could work only because elements of the preexisting neighborhood—the church, the P-patch, and the mixed-use quality—had drawn early twenty-first-century hipsters to the area. With the area finally recognized as an authentic neighborhood in the post-Commons era, Cascade's surviving cultural amenities ironically became a focus and a selling point for developers. "The Cascade neighborhood is a merging of lifestyles," trumpeted the Alcyone promotional literature. "Walking around the neighborhood it's typical to encounter more variety: glassblowers working on their masterpieces; outdoor enthusiasts carrying newly purchased canoes; and St. Spiridon's annual block party with Russian music and cuisine. It's an active neighborhood that takes on many different personalities, all of which collide into an emergence of life." The mixed-use neighborhood that early activists appreciated in Cascade still resonated in literature used to sell the neighborhood in the present. This collision was real and ongoing. The new private ventures of developers—especially the new construction and gentrification of the neighborhood—often clashed with the more publicly minded notions of community found at the P-patch or community center. At the same time, these private and public landscapes blended to make the very "variety" and authenticity that would draw customers. One representative of a prodevelopment group argued that Cascade and south Lake Union would become a competitor in the "world of biotechnology" if the neighborhood succeeded in maintaining itself as a "funky, artistic, sustainable community that lets scientists bump into each other at coffee shops, grocery stores or while dropping off the kids at day care and exchanging ideas,"[53] a kind of "corporate Ecotopia," as the *Seattle Weekly* described it.[54]

When Frank Chopp built his geodesic dome in a Cascade church parking lot during the 1970s, the notion of a place like the Alcyone apartments—with its LEED rating and rooftop gardens—would have been on the wild edge of ecotopian fantasy. And when activists in the 1970s thought about "the Man" or the image of exploitative capitalism, it was usually as a faceless bureaucrat or the all-too-familiar downtown crowd of bankers, department store owners, and hoteliers who ruled many cities in the 1960s. Paul Allen, however, is harder to see in this vein, even though he actually is *the* man. Allen is the only man, in fact, with the predominant control over what happens in, as the *Wall Street Journal* described it, "one of the biggest private urban-renewal projects ever in the U.S."[55]

While no utopia, Allen's planned development shares many ecotopian elements that emerged in the 1960s, realizing the urban village that consensus could not. And Allen employs New Urbanism's ideal typologies born in part from the neighborhood politics of the 1970s. A central aspect of this typology of sustainability is the garden, the most authentic kind of Seattle commons, after all. It makes sense that a product of the 1960s, a baby boomer, would be the mover and shaker behind Cascade and south Lake Union development. His rooftop gardens at the Alcyone Apartments incorporate the familiar elements—a sense of self-sufficiency, community, private piety, and connection to nature—into the private domestic heart of the city. The Alcyone pea-patches[56] suggest a choice of public or private spaces, safely above the street. Since the 1960s, home and garden and all their associations have become central to this reinvention of the urban form, a Seattle brand. Such ideas, a combination of public space, organic produce, and cosmopolitanism, have become selling points for south Lake Union and every other urban neighborhood across the country that offers a Saturday farmers' market or creates a communal garden as part of its community revival efforts. Today, the gardens in Cascade, developers hope, will draw young professionals and a dynamic population of scientists and biotech workers to this up-and-coming neighborhood at Seattle's conflux. Gardens and the accoutrements of bohemian urbanity are part of the potent package, icons of the public in a privatized consumer landscape.

Pursuing something quite different from this private landscape, Jody Aliesan, Tilth's Mark Musick, and other earlier activists hoped to connect the private realm to the public sphere as they challenged received ideas about gender, justice, and consumption, but their projects always risked a

A view of the Alcyone Apartments from the Cascade Community P-Patch, 2010.
Photograph by Jeffrey Sanders.

certain co-optation. According to one historian writing of the period, such "revolutionaries may have challenged the consumer model of the American family," but they could not ultimately "untie the knot binding the consumption of nature to the creation of gender," or for that matter the shape of an urban environment defined by power and class.[57] Certainly Jody Aliesan, neighborhood defenders such as Bob Santos, and other longtime civil rights and poverty activists were fully aware of such conundrums when, with the best intentions, they built urban homesteads and gardens. Unlike the large public-minded gestures of the Commons, however, the home and the garden, with their private connotations, could powerfully reinforce tendencies of retreat in American society. Each group—from Model Cities activists and urban homesteaders to Commons supporters—employed naturalized and often unquestioned ideals of home in their divergent arguments about building communities that were more natural, healthy, and democratic—even sustainable. These discussions about home and nature remind us how the shifting and contested spatial contexts and political contests of the city have shaped debates about environmentalism over the past forty years. Adopting the metaphors of home and garden has not solved the problem of our relationship to the city or to nature; if anything, the notions of home and garden require much more interrogation, especially in an era of global environmental awareness.

Nonetheless, since the 1990s the backyard—or now rooftop—garden (rather than, e.g., the idea of wilderness) has emerged as a fruitful and powerful way for thinking about the human relationship to the nonhuman world, a way to define a politics of responsibility and limits.[58] Yet the concepts of home and garden in the city, as the Commons outcome suggests, may be as troublesome as that of wilderness if not carefully considered.[59] These metaphors were quite familiar to urban activists in Seattle and nationwide as early as the 1970s. Pike Place Market preservationists, Model Cities neighborhood activists, and counterculture food activists helped to cultivate just such a sense of home and garden as a way to connect both the built environment to nature and concerns for nature to their bodies and to social justice.

Allen and his plans for south Lake Union, therefore, force us to confront the contradictions in our formulations of sustainability as a matter of private virtue rather than public process. If Seattleites prefer the small-scale garden, the essentially privatized and intimate spaces that the garden and private acts of consumption represent, then perhaps a certain political style has resulted from the spaces they prefer. This style of politics may ultimately encourage private dealings over public process and localism (even NIMBYism) at the expense of broad collective action. For many opponents of the Commons plan, the scale of the project, unlike the humble community garden, presented an abstraction of urbanism and of nature, something not "sustainable."[60] They were more comfortable in the garden. In addition, these activists fought the plan because they thought that developing south Lake Union would divert much needed public money away from neighborhoods around Seattle. This too is a remnant of the power of neighborhoods (and home) asserted in the 1970s, a reorientation of urban politics that, though certainly warranted in the 1960s, may reinforce a distrust of public process. At its worst, this reorientation echoed nationwide trends of political disengagement during the last thirty years.[61] At its best, it has brought the critical perspective of social justice and equity to grandiose proposals. The seemingly innocent notion of the backyard garden is a fraught metaphor if it ignores this complicated history and politics of place making in cities.

As did earlier ecotopians who showed the disparities within cities, as well as connections between cities and hinterlands in the 1970s, the contemporary movement tries to reveal relationships of power and make the consequences of private consumption in comfortable places manifest on a global scale. The WTO protesters in Seattle at the end of the last century tried to draw us out of our gardens. Beginning in the 1960s a variety of urban actors began an enduring struggle to harmonize nature and the city, to balance public and private control of urban spaces, to make consumer choices political, and to make cities socially just at the same time. By the early twenty-first century, these often competing desires added up to a still elusive and debated idea of "sustainability." Enduring questions about power, about place, and about nature will still be answered on city streets.

Notes

Prologue: "The Battle in Seattle"

1. For a narrative description of the WTO protests from a participant, see Janet Thomas, *The Battle in Seattle: The Story behind the WTO Demonstration* (Golden, Colo.: Fulcrum, 2000). For a complete list of the hundreds of organizations involved, see the archive and collection of essays regarding the World Trade Organization protests in Seattle entitled the WTO History Project, at http://depts.washington.edu/wtohist/. Carl Abbott, in *How Cities Won the West* (Albuquerque: University of New Mexico Press, 2007), 237–38, defines this era and the WTO as indicative of a growing transnational urbanism in which Seattle and other western cities became increasingly integrated into a global network of cities; see also Fred Moody's discussion of the WTO events and Mayor Schell in his memoir *Seattle and the Demons of Ambition: From Boom to Bust in the Number One City of the Future* (New York: St. Martin's, 2003), 1–11.

2. "For Seattle, Triumph and Protest," *New York Times*, October 13.

3. "Trade Pacts Must Safeguard Workers, Union Chief Says," *New York Times*, November 20, 1999.

4. Timothy Egan, "Black Masks Lead to Pointed Fingers in Seattle," *New York Times*, December 2, 1999.

5. "World Trade Groups Inspire Protest," *New York Times*, November 29, 1999.

6. WTO History Project.

7. "Trade Groups Inspire Protest."

8. Andres R. Edwards, *The Sustainability Revolution: Portrait of a Paradigm Shift* (Gabriola Island, B.C., Canada: New Society, 2005), 15–17. The Millennium Summit in 2000 affirmed this notion of sustainability and defined the "Millennium Goals," which include the environmental sustainability goals; see http://www.un.org/millenniumgoals/environ.shtml.

9. My use of *sustainability* draws from the United Nations' evolving language and definition, which emphasizes social justice as a key component, but I am interested in showing how the earlier urban environmental movement of the mid-1960s set the critical groundwork for these contemporary ideas and practices. For a brief history of recent developments in defining sustainability, see Edwards, *Sustainability Revolution*, 15–17; see also World Commission on Environment and Development, *Our Common Future* (Oxford: Oxford University Press, 1987), also known as the Brundtland Report, issued by the

239

U.N. commission in the late 1980s following its initial 1983 meeting in Stockholm. The Stockholm meeting put environment and sustainability on the world body's agenda for the first time. By the early 1990s, the Rio Conference had further emphasized social justice as part of sustainability. The preamble to Agenda 21—the conference's set of goals—describes the United Nations' vision for "a global partnership for sustainable development"; see Daniel Sitarz, ed., *Agenda 21: The Earth Summit Strategy to Save Our Planet* (Boulder, Colo.: Earthpress, 1993), 29; for the full Rio Declaration on Environment and Development principles, see http://www.un.org/esa/dsd/agenda21/res_agenda21_00 .shtml.

10. "Trade Groups Inspire Protest."

11. WTO History Project; The "Battle in Seattle," a term the press used to describe the 1999 events, has a deeper resonance. The "Battle of Seattle" was a January 26, 1856, clash between settlers in the area that is now Seattle and Indians of the Puget Sound region; see Coll-Peter Thrush, *Native Seattle: Histories from the Crossing-Over Place* (Seattle: University of Washington Press, 2007). Seattle also had a deeper history of political dissent and radicalism; see Dana Frank, *Purchasing Power: Consumer Organizing, Gender, and the Seattle Labor Movement, 1919-1929* (Cambridge: Cambridge University Press, 1994); Albert A. Acena, "The Washington Commonwealth Federation: Reform Politics and the Popular Front," Ph.D. diss., University of Washington, 1975. The recent street protests and organized actions against the World Trade Organization in Seattle brought this history of radical possibility back to the surface.

12. "Handling Protest in Seattle," *New York Times*, December 4, 1999.

13. "Black Masks"; for more on the aftermath of the protest and the findings of an accountability panel, see the Citizens' Panel on WTO Operations' report to the Seattle City Council WTO Accountability Committee at http://www.seattle.gov/wtocommittee/ panel3final.pdf.

14. "Seattle Is Stung, Angry and Chagrined as Opportunity Turns to Chaos," *New York Times*, December 2, 1999.

15. Steven Greenhouse, "World Trade Group Inspires Protesters," *New York Times*, November 29, 1999.

16. The ELF traces its origins to Great Britain. Seattle and its surrounding suburbs, like many sprawling and booming areas in the West, stood out as a target for activists who saw the ill-effects and hypocrisy of a wealthy populace who claimed to love nature but allowed urban sprawl into wild areas and the proliferation of giant sport utility vehicles during the 1990s; for an excellent discussion of the Earth Liberation Front, their recent history in the Pacific Northwest, and anarchist groups in Oregon, see Bruce Barcott, "From Tree Hugger to Terrorist," *New York Times*, April 7, 2002. For perhaps a century, and certainly over the last thirty years, Seattle and Portland have been rivals, competing for "most livable" city status, or more recently most "sustainable." Portland has succeeded in many ways where Seattle has not in terms of mass transit, open space, and significant restrictions on urban sprawl; see Abbott, *Metropolitan Frontier*, 145–47. Seattle has a slightly larger population, but it covers a smaller area (83 square miles to 134 for Portland), making its population density almost twice that of Portland (6,717.2 people per square mile vs. 3,939.3). Seattle, then, with a smaller area and larger suburban areas, is comparatively more diverse and complex than the city to the south. For a full comparison of size, diversity, and standard of living for the two cities, see census data for Seattle at http://quickfacts.census.gov/qfd/states/53/5363000.html and for Portland at http://quickfacts.census.gov/qfd/states/41/4159000.html. In 2009, according to the Natural Resource Defense Council, Seattle rated first among the most sustainable cities, with San Francisco second and Portland third. The three ecotopian cities—not including

the progressive Vancouver, B.C.—have each rated high in these annual evaluations over the last several years. In each case, the issue of affordability, or standard of living, was included; see http://smartercities.nrdc.org/rankings/large.

17. By 1999, the contradictions—especially over the connections among race, class, and environmental politics—had yet to be resolved. The caricature of the 1960s has not been helpful in thinking historically about social changes or eruptive events such as the WTO protests. However, historians of the 1960s "revolutions" have recently worked to place the sudden and explosive changes of that period within a much broader framework of postwar American history. David Farber, for instance, argues that what seemed like sudden changes were part the "changing nature of cultural authority and political legitimacy" in a period of flux between World War II and the 1970s; see Farber, *The Sixties: From Memory to History* (Chapel Hill: University of North Carolina Press, 1994), 1. For another example of this effort to "ground most of the explosive events of the 1960s" in history, see Farber, *The Age of Great Dreams: America in the 1960s* (New York: Hill and Wang, 1994), 3. Thomas Sugrue, *The Origins of the Urban Crisis: Race and Inequality in Postwar Detroit* (Princeton, N.J.: Princeton University Press, 1996), and other works on specific aspects of this period have similarly begun to place the period's tumult into better and deeper context.

18. Cultural geographers make this connection between the production of physical and social space and crises of capitalism, with the late 1960s constituting one of these moments of rupture; see, e.g., David Harvey, *The Condition of Postmodernity* (Oxford: Blackwell, 1990); Edward Soja, *Postmodern Geographies: The Reassertion of Space in Critical Social Theory* (London: Verso, 1989).

19. Robert Gottlieb has done pioneering work in placing the history of environmental activism in the urban context as well as showing the movement's more diverse ethnic, gender, and class roots in *Forcing the Spring: The Transformation of the American Environmental Movement* (Washington, D.C.: Island, 1993).

20. Like the politics of race and class at midcentury, environmental activism shares the same spatial framework that scholars use the "urban crisis," suburbanization, desegregation, and "white flight" have increasingly applied to emerging patterns of metropolitan politics and culture. For a collection that represents this approach to the spatial politics of the era, see Kevin Michael Kruse and Thomas J. Sugrue, *The New Suburban History* (Chicago: University of Chicago Press, 2006). Robert O. Self's work on Oakland is particularly helpful in linking urban grassroots political movements of the inner city during the late 1960s and early 1970s to the spatial politics of the metropolitan era, showing how local actors were brought to a heightened awareness of the production of space in their local environments; see Self, *American Babylon: Race and the Struggle for Postwar Oakland* (Princeton, N.J.: Princeton University Press, 2003).

21. With the exception of Robert Gottlieb's *Forcing the Spring*, the dominant narrative of environmental history, and the role of activists in that history, does little to contextualize grassroots environmental activism or connect it to other movements for social change, particularly to urban places and people. For examples of the narrative, see Hal Rothman, Gerald D. Nash, and Richard W. Etulain, *The Greening of a Nation? Environmentalism in the United States since 1945* (Fort Worth, Tex.: Harcourt Brace, 1998); Kirkpatrick Sale, *The Green Revolution: The American Environmental Movement, 1962–1992* (New York: Hill and Wang, 1993); Theodore Steinberg, *Down to Earth: Nature's Role in American History* (Oxford: Oxford University Press, 2002). Adam Rome has argued for a better evaluation of the role played by the "growing discontent of middle class women," for example, and a better connection of environmentalism to other political movements in the narrative of the 1960s. He suggests such an approach might re-

veal elements that "contributed in key ways to the emergence of environmentalism"; see Rome, "'Give Earth a Chance': The Environmental Movement and the Sixties," *Journal of American History* 90, no. 2 (Sept. 2003): 525–54; see also Rome's discussion of suburban homeowners and the rise of the environmental movement in *The Bulldozer in the Countryside: Suburban Sprawl and the Rise of American Environmentalism* (New York: Cambridge University Press, 2001). For an informative recent collection that helps fill in this context, as well as the variety of approaches and actors in this more variegated story of environmentalism, see Michael Egan and Jeff Crane, *Natural Protest: Essays on the History of American Environmentalism* (New York: Routledge, 2009); for a recent example of scholarship that links the African American experience to environmental thought and activism antedating the environmental justice movement, see Kimberly K. Smith, *African American Environmental Thought: Foundations* (Lawrence: University Press of Kansas, 2007).

22. For more on this history and the influence of urban renewal on downtown living, see Paul Erling Groth, *Living Downtown: A History of Residential Hotels in the United States* (Berkeley: University of California Press, 1994).

23. Seattleites were not alone in their concern for the built and natural environments of cities. William Issel explores this important terrain with regard to freeway protests in San Francisco and the emergence of a new kind of liberalism and environmentalism in "Land Values, Human Values, and the Preservation of the 'City's Treasured Appearance': Environmentalism, Politics, and the San Francisco Freeway Revolt," *Pacific Historical Review* 68 (1999): 611–46; see also Richard DeLuca, *"We The People!" Bay Area Activism in the 1960s* (San Bernardino, Calif.: Borgo, 1994); Richard Walker, *The Country in the City: The Greening of the San Francisco Bay Area* (Seattle: University of Washington Press, 2007).

24. For general discussions of works that seek to establish links between environmental and urban history, see Timothy J. Gilfoyle, "White Cities, Linguistic Turns, and Disneylands: The New Paradigm of Urban History," *Reviews in American History* 26, no. 1 (1998): 175–204; Martin Melosi, "The Place of the City in Environmental History," *Environmental History Review* 17 (Spring 1993): 1–23; Raymond A. Mohl, "City and Region: The Missing Dimension in U.S. Urban History," *Journal of Urban History* 25 (Fall 1998): 3–21; Joel A. Tarr, "Urban History and Environmental History in the United States: Complementary and Overlapping Fields," in *Environmental Problems in European Cities of the Nineteenth and Twentieth Century,* ed. Christoph Bernhardt (New York: Waxmann, Meunster, forthcoming). Yet few recent studies place environmental activism within these overlapping contexts in the 1960s and after. Dolores Hayden's approach to urban space—one that attempts to include cultural, social, and even natural spaces as part of her discussion of contested place making—has been helpful for scholars interested in these overlapping aspects of urban environments. When grassroots actors spoke of the built landscape, they often used organic terms. And sometimes when actors spoke of nature, they were actually speaking of social concerns; see Hayden, *The Power of Place* (Cambridge, Mass.: MIT Press, 1997), 1–40. Hayden draws from Henri Lefebvre in conceiving of the social production of space; see Lefebvre, *The Production of Space* (Cambridge, Mass.: Basil Blackwell, 1991). Lefebvre, although he omits an analysis of nature from his work, has been helpful for environmental historians, who find his ideas particularly applicable for understanding how different spaces—including ones deemed "natural"—are shaped as part of social processes. For a similar discussion of the intersection of nature and socially produced spaces of meaning, see Richard White, *The Organic Machine* (New York: Hill and Wang, 1995); White, "The Nationalization of Nature," *The Nation and Beyond,* special issue, *Journal of American History* 86, no. 3 (Dec. 1999): 976–

86. See also William Cronon, "Kennecott Journey: The Paths out of Town," in *Under an Open Sky: Rethinking American Western Past*, ed. William Cronon, George Miles, and Jay Gitlin (New York: W. W. Norton, 1992), 28–51; and Matthew W. Klingle, *Emerald City: An Environmental History of Seattle* (New Haven, Conn,: Yale University Press, 2007), the most thorough recent work on environmental history in Seattle. The direction of many of these histories, in the vein of earlier historians of vernacular and women's space, is away from a discussion of nature in terms of pristine wilderness and toward an understanding of many landscapes—including urban landscapes and domestic spaces—shaped by natural and cultural forces.

25. Although Lizabeth Cohen does not engage the environmental movement in any detail in her work, her definition of "citizen consumer" is useful to understanding how Seattleites began to see their consumer choices and private decisions as part of civic life; see Cohen, *Consumer's Republic: The Politics of Mass Consumption in Postwar America* (New York: Alfred A. Knopf, 2003), 1–18; for arguments about the shaping of the urban form and specifically the way consumption shaped the immediate postwar landscape, see Cohen, "Is There an Urban History of Consumption?" *Journal of Urban History* 29 (Jan. 2003): 87–106. Cohen makes interesting links between suburbanization in the postwar period and the deterioration and change—both economically and in physical appearance and form—in the inner city. For more direct links between the history and politics of consumption and environmental history, see Matthew W. Klingle, "Spaces of Consumption in Environmental History," *History and Theory* 42, no. 4 (2003): 94–110.

26. Linda Nash, for instance, has helped to redirect the discussion of what constitutes environment, or the terrain of environmental concern and activism, toward the human body. In her work, which explores a history of bodies and places in California long before and then during the environmental justice movement, she explains how bodies, places, and politics became entwined by the 1960s; see Nash, *Inescapable Ecologies: A History of Environment, Disease, and Knowledge* (Berkeley: University of California Press, 2006). In Seattle's Model Cities neighborhood during the late 1960s activists and city workers began to trace similar connections. See also Gregg Mitman, "In Search of Health: Landscape and Disease in American Environmental History," *Environmental History* 10, no. 2 (Apr. 2005): 184–210.

27. After the 1960s, women's efforts to solve urban health problems or to define new urban green space and housing options, redefine the domestic sphere, or preserve places from the ravages of urban renewal—often in opposition to predominantly male power structures and engineers in the city—provided a crucial thread in the story of ecotopia's origins. See Gottlieb, *Forcing the Spring*, for more on this history. For a discussion of the way historians have framed women's urban environmental activism, especially in the early 1960s, see Lewis L. Gould, *Lady Bird Johnson and the Environment* (Lawrence: University Press of Kansas, 1988), 1–6; for a discussion of gender as a framework for understanding how people have acted on and understood nature, see Virginia Scharff, "Man and Nature! Sex Secrets of Environmental History," Maril Hazlett, "Voices from the Spring: *Silent Spring* and the Ecological Turn in American Health," and Amy Green, "She Touched Fifty Million Lives": Gene Stratton-Porter and Nature Conservation," in *Seeing Nature through Gender*, ed. Virginia Scharff (Lawrence: University Press of Kansas, 2003), 3–19, 103–28, 221–41.

28. John Findlay describes the specific shape of midcentury utopianism in Seattle when it was displayed at the 1962 world's fair—a utopian, space-age world's fair that celebrated Seattle's role in the aeronautics and defense industry; see Findlay, *Magic Lands: Western Cityscapes and American Culture after 1940* (Berkeley: University of California Press, 1992). Findlay argues that such ideas gave shape to postwar urbanism. Although

fairs such as Century 21 did a great deal to structure postwar urbanism, Seattle activists in the late 1960s and 1970s emerged as a city in reaction to such urban visions. In the wake of projects such as the fair and urban renewal projects of the period, people in Seattle and other cities began to reject large-scale plans and instead nurture overlapping concerns for the preservation of both built and natural landscapes.

29. Andrew Kirk makes similar arguments about the politics of green consumption in *Counterculture Green: The Whole Earth Catalog and American Environmentalism* (Lawrence: University Press of Kansas, 2008); these activists created the early infrastructure for what has become the contemporary "locovore" movement. For a discussion of this history and this contemporary concern about nearby food, see Michael Pollan, *The Omnivore's Dilemma: A Natural History of Four Meals* (New York: Penguin, 2006). For other examples of the regional and appropriate technology focus of the counterculture, see Kirk, "Appropriating Technology: *The Whole Earth Catalog* and Counterculture Environmental Politics," *Environmental History* 6, no. 2 (Apr. 2001): 374–94; idem, "'Machines of Loving Grace': Alternative Technology, Environment, and the Counterculture," in *Imagine Nation: The American Counterculture of the 1960s and '70s*, ed. Peter Braunstein and Michael William Doyle (New York: Routledge, 2002), 353–78; Jordan Kleiman, "The Appropriate Technology Movement in American Political Culture," Ph.D. diss., University of Rochester, 2001; idem, "Local Food and the Problem of Public Authority," *Technology and Culture* 50, no. 2 (2009): 399–417. For an discussion of the way nature, gender, and food practices are interrelated, see Douglas C. Sackman, "Putting Gender on the Table: Food and the Family Life of Nature," in *Seeing Nature through Gender*, ed. Virginia Scharff (Lawrence: University Press of Kansas, 2003), 169–93.

30. The term *counterlandscape* borrows from Michel Foucault's notion of "heterotopias"; see "Of Other Spaces—'Heterotopias' and 'Panopticon,'" in *Rethinking Architecture*, ed. Neil Leach (New York: Routledge, 1997), 350–67.

31. Both these terms—*region* and *sustainability*—are contested. Definitions of sustainability have ranged widely. I offer the United Nations' definition, but in the past ideas of sustainability usually involved a sense of self-sufficiency that included local energy sources, local food networks, and practices that had low impact on the environment. These ideas or desires rarely manifested in reality. For an instructive deconstruction of the idea of region in the Pacific Northwest, see John Findlay, "A Fishy Proposition: Regional Identity in the Pacific Northwest," in *Many Wests: Place, Culture, and Regional Identity*, ed. David M. Wrobel and Michael Steiner (Lawrence: University Press of Kansas, 1997), 37–70.

32. For the history of domestic space and architecture, see Dolores Hayden, *Redesigning the American Dream: Gender, Housing, and Family Life* (New York: W. W. Norton, 2002); idem, *The Grand Domestic Revolution: A History of Feminist Designs for American Homes, Neighborhoods, and Cities* (Cambridge, Mass.: MIT Press, 1981); Gwendolyn Wright, *Building the Dream: A Social History of Housing in America* (Cambridge, Mass.: MIT Press, 1983). For a more general work on vernacular architecture, see J. B. Jackson, *A Sense of Place, a Sense of Time* (New Haven, Conn.: Yale University Press, 1994).

33. During the 1990s, Seattle became one of the promising examples of New Urbanism and the urban village concept of planning. For a portrait of the way many Seattle planners saw the city in the 1990s, see Douglas Kelbaugh, *Common Place: Toward Neighborhood and Regional Design* (Seattle: University of Washington Press, 1997); idem, *Repairing the American Metropolis* (Seattle: University of Washington Press, 2002).

34. My discussion of the sustainable city echoes a recent article by David Owen, who argued in the *New Yorker* that Manhattan, despite common American perceptions, is the most "sustainable" kind of environment because of its concentration of people,

resources, and energy; see Owen, "Green Manhattan," *New Yorker*, October 18, 2004, 111–23.

35. Randy Hayes, "Protests with Purpose," *New York Times*, December 12, 1999.

Chapter 1: Market

1. Mark Tobey, *Mark Tobey: The World of a Market* (Seattle: University of Washington Press, 1964) (quotation). For more on Mark Tobey and the art scene in early twentieth-century Seattle, see Roger Sale, *Seattle, Past to Present: An Interpretation of the History of the Foremost City in the Pacific Northwest* (Seattle: University of Washington Press, 1976), 166–72; Sheryl Conkelton and Laura Landau, *Northwest Mythologies: The Interactions of Mark Tobey, Morris Graves, Kenneth Callahan, and Guy Anderson* (Seattle: Tacoma Art Museum, in association with University of Washington Press, 2003).

2. Tobey, *Mark Tobey*.

3. "Joint Release, Seattle City Planning Commission and the Central Association of Seattle," February 1963, Office of the Mayor, external correspondence, Central Association of Seattle, box 20, fol. 5, Seattle Municipal Archives (SMA).

4. For examples of Ehrlichman's attitudes about the future of the city, see "Central Group Elects Ehrlichman," *Seattle Times*, May 17, 1963; "Remarks by Ben B. Ehrlichman, Central Association Annual Meeting," May 18, 1964, Office of the Mayor, external correspondence, Central Association of Seattle, box 54, fol. 9, SMA. For more on the fair and how Seattle's hopes and approach to modernization reflected broader patterns in the western United States, see John Findlay, *Magic Lands: Western Cityscapes and American Culture after 1940* (Berkeley: University of California Press, 1992), 214–64.

5. Tobey, *Mark Tobey*; "Pike Plaza Redevelopment: A Preliminary Feasibility Study," August 1964, University of Washington Library, Special Collections (UWL-SC). For the famous description of the era's postscarcity thinking, see John Kenneth Galbraith, *The Affluent Society* (Boston: Houghton Mifflin, 1958), 1–5.

6. Long before the latest iteration of this sensibility—the "locavore" movement—consumers in Seattle took local food for granted, though many were upset when that local landscape began to disappear.

7. "Bara Bara: Scattered," 1981, Pike Place Market Visual Images and Audiotapes, Department of Community Development, box 9, fol. 81, SMA; Nikkei farming operations in the Northwest encompassed both larger operations south of the city, as far away as eastern Washington and as close by as Bainbridge and Vashon Islands, and the many smaller truck farms around Seattle's immediate city boundaries and on land that is now part of its suburbs (see John Adrian Rademaker, "The Ecological Position of the Japanese Farmers in the State of Washington," Ph.D. diss., University of Washington, 1939, 178–92); "This Market Is Yours," Pike Place Market Preservation and Development Authority: Historical/Biographical, Department of Community Development, box 133, fol. 1, SMA; see also Judy Mattivi Morley, *Historic Preservation and the Imagined West* (Lawrence: University Press of Kansas, 2006), 91–125; Alice Shorett and Murray Morgan, *The Pike Place Market: People, Politics, and Produce* (Seattle: Pacific Search Press, 1982).

8. "1919 Plan for Stalls" and "The New Plan, 1922," plans for the Pike Place Market alterations and construction, 1910–1939, City of Seattle, Department of Buildings, Department of Administrative Services, box 8, fol. 5, SMA.

9. "This Market Is Yours."

10. "Bara Bara: Scattered." For more on this period in Seattle, see Monica Sone, *Nisei Daughter* (Seattle: University of Washington Press, 1979); for an excellent discussion of Seattle's *Nihonmachi* and the experience of World War II, see Gail Dubrow and Donna Graves, *Sento at Sixth and Main* (Washington, D.C.: Smithsonian Institution, 2004).

11. "This Market Is Yours."

12. "Bara Bara: Scattered."

13. For an excellent discussion of this period and the issue of returning Japanese Americans in the Puget Sound, see Jennifer Speidel, "After Internment: Seattle's Debate over Japanese Americans' Right to Return Home," Seattle Civil Rights and Labor History Project, http://depts.washington.edu/civilr/after_internment.htm.

14. "Bara Bara: Scattered."

15. For a thorough recent overview of metropolitanization in the United States, see Dolores Hayden, *Building Suburbia: Green Fields and Urban Growth, 1820–2000* (New York: Vintage, 2003), 3–18, 128–53.

16. Lizabeth Cohen, *Consumer's Republic: The Politics of Mass Consumption in Postwar America* (New York: Alfred A. Knopf, 2003), 195. For a general discussion of this era in western development, see Carl Abbott, *Metropolitan Frontier: Cities in the Modern American West* (Tucson: University of Arizona Press, 1993); for discussion of Boeing during this period, see T. M. Sell, *Wings of Power: Boeing and the Politics of Growth in the Northwest* (Seattle: University of Washington Press, 2001). Cohen describes this era of rapid development in *Consumer's Republic*, 1–18; for arguments about the shaping of the urban form and specifically on the way consumption shaped the postwar landscape, see Cohen, "Is There an Urban History of Consumption?" *Journal of Urban History* 29 (Jan. 2003): 87–106.

17. Matthew Klingle, *Emerald City: An Environmental History of Seattle* (New Haven, Conn.: Yale University Press, 2007), 205–8.

18. The years between 1954 and 1959 saw the most dramatic drop in the number of farms in King County, from 5,181 to 2,952. Between these years acreage fell by 20.94 percent; see "Saving Farming and Open Space: The Report of the Citizens Study Committee to the Executive and Council of King County, King County (Wash.)," Save Our Local Farmlands Committee, Citizens Study Committee, 1979, 47, UWL-SC (these statistics are based on the 1965 agricultural census).

19. See Abbott, *Metropolitan Frontier*, 64–78; Sale, *Seattle*, 173–215.

20. Charles Russell, "Kent: Fast Beating 'Heart of the Valley of Plenty,'" pictorial, *Seattle Post-Intelligencer*, September 2, 1956.

21. Ibid.

22. See "Lake Fenwick Tracts" and many other examples of restrictive covenants for Seattle and King County, "Racial Restrictive Covenants," Seattle Civil Rights and Labor History Project, University of Washington, http://depts.washington.edu/civilr/covenants.htm.

23. Russell, "Kent," 4–5.

24. "Time for Boeing to Wake Up," *Seattle Argus*, May 3, 1963.

25. "Boeing Invests Heavily in Seattle Area Facilities," *Seattle Times*, March 13, 1963.

26. "As Goes Boeing, So Goes Seattle," *Seattle Times*, March 31, 1963.

27. See "Boeing Will Spend Millions on Kent Research Center," *Seattle Times*, February 4, 1964; "Boeing to Build $15 Million Kent Space Center, *Seattle Post-Intelligencer*, February 5, 1964.

28. "Boeing to Build."

29. "US Aid for Expansion," *Seattle Times*, May 26, 1966.

30. Ibid.

31. "Now It's Green (back) River Valley," *Seattle Times*, October 24, 1965.

32. Ibid.

33. "Rocket Research to Build New Plant in Kent Area," *Seattle Times*, January 16, 1966.

34. "Boomtown Northwest: Kent's Growth Hits Staggering Proportions," *Seattle Argus*, April 15, 1968.

35. Ted Steinberg describes this exurban farming landscape in *Down to Earth: Nature's Role in American History* (New York: Oxford University Press, 2002), 162–63, 178–79, 189.

36. Cassandra Tate, "Ehrlichman, Ben Bernard (1895–1971)," Historylink, www.historylink.org; Klingle, *Emerald City*, 205.

37. One of the first public discussions of these trepidations about planning for a metropolitan future and its effect on downtown could be found in a joint policy statement published by the City of Seattle and the newly created Central Association of Seattle, "Planning the Future of Seattle's Central Area," Seattle Planning Commission, 1959, UWL-SC.

38. See Allison Isenberg, *Downtown America: A History of the Place and the People Who Made It* (Chicago: University of Chicago Press, 2004). The Central Association's annual reports between 1960 and 1970, as well as the newsletter published by the Chamber of Commerce, *Seattle Business*, show ample evidence that the Seattle planning community tapped into a much broader national effort to revitalize cities throughout the United States and a planning culture that shared their assumptions about the city and about consumers. The Central Association also helped to organize urban revitalization tours in other cities during the 1960s for its members. The 1963–64 annual report, titled "Downtown Redevelopment: Seattle's Challenge of the Decade" (Central Association of Seattle, 1963, UWL-SC), for instance, supported local plans with snapshots of similar redevelopment plans in Fresno, California; St. Louis, Missouri; and Hartford, Connecticut; see also the newsletter *Seattle Business*, 1959–70, UWL-SC.

39. "Holding Back the Waters," *Seattle Times*, May 6, 1965.

40. "Tornado's Path? No—It's the Freeway Route," *Seattle Times*, March 22, 1959.

41. John Findlay makes this argument about the fairgrounds in *Magic Lands*, 214–18. Robert Self challenges this idea of "white flight," arguing that the term *flight* does not accurately describe a process in which middle-class whites were drawn by subsidies and incentives to a suburban landscape; see Self, *American Babylon: Race and the Struggle for Postwar Oakland* (Princeton, N.J.: Princeton University Press, 2003), 1–20.

42. "Planning the Future," 5, UWL-SC.

43. "Progress Report," Sunday supplement, *Seattle Times*, October 29, 1961.

44. "Balanced Transportation Plan for City Stressed," ibid.

45. "Downtown for People," ibid.

46. "Joint Release," Seattle City Planning Commission and Central Association of Seattle.

47. Ibid.

48. "Seattle Plans Development," Seattle City Planning Commission and Central Association of Seattle, February 1963, Office of the Mayor, external correspondence, box 20, fol. 5, SMA.

49. Ibid.

50. *Pike Plaza Redevelopment Project*, pamphlet, UWL-SC.

51. "Joint Release."

52. "Pike Plaza Redevelopment," August 1964, UWL-SC.

53. Ibid.

54. "Central Group Elects Ehrlichman," *Seattle Times*, May 17, 1963; Ben B. Ehrlichman, "Grasp a Thistle Firmly . . . ," *Central Association of Seattle Annual Report, 1963–64*, 1964, UWL-SC.

55. Ehrlichman, "Grasp a Thistle."

56. "Remarks by Ben B. Ehrlichman, Central Association Annual Meeting," May 18,

1964, Office of the Mayor, external correspondence, Central Association of Seattle, box 54, fol. 9, UWL-SC.

57. Judy Mattivi Morley (*Historic Preservation*, 124–26) provides an excellent discussion of the market's meaning for Seattleites as well as the role of postwar modes of consumption in shaping the market as a tourist attraction that ultimately depends, ironically, on the public sector to perpetuate its ideal image as a more "honest" expression of small-scale farming. Of course, the postwar suburban landscape that replaced the old farm fields near cities such as Seattle were just as heavily subsidized beginning soon after the war.

58. Architectural historian Jeffrey Ochsner describes the evolution of Steinbrueck's thought, aesthetic concerns, and interest in social issues in "Victor Steinbrueck Finds His Voice: From *The Argus* to *Seattle Cityscape*," *Pacific Northwest Quarterly* 99, no. 3 (Summer 2008): 122–33; Ochsner emphasizes Steinbrueck's love of nature and commitment to protecting the vernacular landscape from the ravages of freeway development beginning in the 1950s as catalysts for his activism.

59. To influence the public in support of their efforts, for example, Allied Arts created an annual election survey that evaluated the city council on various urban improvement issues. In 1967 the survey's first question was, "Do you feel that the elimination of urban ugliness and enhancement of beauty have a place in the program of a well-governed city?" ("Allied Arts Election Survey: Pocket Guide to Your Voting Booth, November 7, 1967, City Council Election," Allied Arts of Seattle newsletter, 1967, UWL-SC). The survey also queried city council members-to-be on "undergrounding" wires, support for a city tax earmarked for the arts, regulation of high-rises, and zoning ordinances for signage in the city. In the late 1960s and early 1970s, these new activist civic groups would fight the older downtown business establishment to stop the destruction of historic sites such as Pioneer Square and the market, which planning would have replaced with new parking lots. For a national context and history of this period, see Bernard J. Frieden and Lynne B. Sagalyn, *Downtown, Inc.: How America Rebuilds Cities* (Cambridge, Mass.: MIT Press, 1990), 171–98.

60. See Sale, *Seattle*, 216–52.

61. "We've Toyed with Seattle Long Enough," *Seattle Magazine*, February 1966, 14–17.

62. Sale, *Seattle*, 223. For more on the fight over the market and the eventual preservation plans, see Morley, *Historic Preservation*; Sohyun Park Lee, "Conflicting Elites and Changing Values: Designing Two Historic Districts in Downtown Seattle, 1958–73," *Planning Perspectives* 16, no. 3: 397–477.

63. Steinbrueck, "Book Reviews," *Seattle Magazine*, 1964, 24–25.

64. Ibid.

65. Individual members of the faculty of the College of Architecture and Urban Planning, University of Washington, to the city council, 1970, Victor Steinbrueck Papers, 1931–86, box 3, fol. 3, UWL-SC.

66. Ochsner, "Victor Steinbrueck."

67. Meyer Wolfe, *Towns, Time, and Regionalism*, Urban Planning and Development Series, no. 2 (Seattle: Dept. of Urban Planning, University of Washington, 1963), i, Washington State University Library (WSUL); Wolfe, *Uses of the Anachronistic* (Seattle: Dept. of Urban Planning, University of Washington, 1970), WSUL. See also Ernst L. Gayden, "The Role of Ecological Theory in Urban Planning," Ph.D. diss., University of Illinois at Chicago, 1967.

68. Victor Steinbrueck, *Seattle Cityscape* (Seattle: University of Washington Press, 1962), 27.

69. Victor Steinbrueck, *Seattle Cityscape #2* (Seattle: University of Washington Press, 1973), 6.

70. Ibid., 28.

71. Steinbrueck, *Seattle Cityscape*, 36.

72. Jane Jacobs, like Steinbrueck, praised the community-nurturing wisdom of maintaining urban areas that developers believed to be "blighted," and she offered a direct critique of the contemporary planning ideas of the early 1960s in *Death and Life of Great American Cities* (New York: Modern Library Edition, 1993), 5.

73. Victor Steinbrueck, "First Avenue is First Avenue is First Avenue . . . ," *Northwest Today, Seattle Post-Intelligencer*, December 5, 1965, 7; Steinbrueck, "Reply to Central Association Regarding Pike Place Market Destruction," Victor Steinbrueck Papers, 1931–86, box 3, fol. 3, UWL-SC.

74. Victor Steinbrueck, *Market Sketchbook* (Seattle: University of Washington Press, 1968).

75. Ibid. Steinbrueck reflected a general postwar movement and echoed similar developments in Europe at the same time—namely, the activities and ideas of the Situationists, who explored the nooks and crannies of postwar Paris with similar relish; see Simon Sadles, *The Situationist City* (Cambridge, Mass.: MIT Press, 1998).

76. See Isenberg, *Downtown America*, 1–2, 166–202.

77. "Habits and Opinions of Seattle's 63,000 Downtown Workers," Central Association of Seattle, 1960, UWL-SC.

78. Ibid.

79. Isenberg, *Downtown America*, 166–202.

80. "Habits and Opinions."

81. "100 Marchers Call for Freeway Lid," *Seattle Times*, June 1, 1961.

82. For an overview of this "seed time" in a coalescing movement in the early 1960s and after, with an emphasis on Rachel Carson's influence as well as groups dominated by women, such as the Audubon Society, see Kirkpatrick Sale, *The Green Revolution: The American Environmental Movement, 1962–1992* (New York: Hill and Wang, 1993), 1–45; another useful overview can be found in Hal Rothman, Gerald D. Nash, and Richard W. Etulain, *The Greening of a Nation? Environmentalism in the United States since 1945* (Fort Worth, Tex.: Harcourt Brace, 1998). See also classic works such as Samuel P. Hays, *Beauty, Health, and Permanence: Environmental Politics in the United States, 1955–1985* (New York: Cambridge University Press, 1987). To gain a sense of the powerful influence of women in this movement, see Lewis L. Gould, *Lady Bird Johnson and the Environment* (Lawrence: University Press of Kansas, 1988); also see Vera Norwood's discussion of women's activism in the 1960s and Rachel Carson's influence in *Made from This Earth: American Women and Nature* (Chapel Hill: University of North Carolina Press, 1993), 143–71. For a recent discussion of what she terms the "ecological turn" in political culture, see Maril Hazlett, "Voices from the Spring: *Silent Spring* and the Ecological Turn in American Health," in *Seeing Nature through Gender*, ed. Virginia Scharff (Lawrence: University Press of Kansas, 2003), 103–28.

83. Many of the same developers and businesspeople in downtown Seattle who hoped to remake the city also supported plans to build rapid transit in the early 1960s as part of their plans for both the commercial district and the region. While the CA planned redevelopment, advocates of good government and regional planning—a group called the Municipal League—created Forward Thrust, a massive proposed $820 million program, the local equivalent of the Great Society; promoting this program, the league would eventually ask voters to approve bonds for the construction of a light rail system similar to BART in San Francisco; funds for freeway construction, community centers, sewer im-

provements, and a multipurpose domed stadium; and over $118,000,000 for parks and open space. The brainchild of Jim Ellis, a Seattle lawyer and civic activist who had led the fight to create METRO, a countywide governmental agency in the late 1950s, Forward Thrust was the regional, countywide planning version of Seattle's neoprogressive dreams—a great, rational plan to accommodate and shape future growth in the area. Voters would ultimately turn down substantial sections of the bond issue, particularly the $385,000,000 for light rail, by 1968. See Klingle, *Emerald City*, 234–35.

84. Memorandum of Councilman Wing Luke enclosing criticism of proposed CBD plan by Victor Steinbrueck, September 30, 1963, comptroller file no. 249268, SMA.

85. Ibid.

86. Petition of citizens for maintenance of the present appearance and general character of the Pike Place Public Market, City Clerk, October 28, 1963, comptroller file no. 249399, SMA.

87. Mrs. Mary Carpenter to city council, petition of Mrs. Mary Carpenter and others re: Pike Place Market, September 4, 1964, comptroller file no. 251647, SMA.

88. Mrs. Robert Ray to city council, ibid.

89. These advocates of local food and fresh vegetables long antedated the current slow food and locovore movement but suggested the beginnings of this movement in places such as Seattle, where people had begun to worry about the city and nature at the same time. For a popular recent discussion, see Michael Pollen, *Omnivore's Dilemma: A Natural History of Four Meals* (New York: Penguin, 2006).

90. Adam Rome, "'Give Earth a Chance': The Environmental Movement and the Sixties," *Journal of American History* 90, no. 2 (Sept. 2003): 525–54 (esp. 527).

91. Victor Steinbrueck, National Register Nomination Inventory Nomination Form, Victor Steinbrueck Papers, 1931–86, box 3, fol. 1, UWL-SC.

92. John H. Tuttle, Ph.D., to city council, petition of Timothy Pettit for preservation of the Pike Place Public Market in its present form, September 8, 1969, comptroller file no. 264592, SMA.

93. University of Washington students to city council, ibid.

94. "Seattle Receives Model Cities Aid of $5.2-Million," *New York Times*, December 24, 1968, 19.

95. R. Duke to city council, petition of Timothy Pettit for preservation of the Pike Place Public Market in its present form, September 8, 1969, comptroller file no. 264592, SMA.

96. Elizabeth Murphy to city council, ibid.

97. Kate M. Baldwin and Stephen Baldwin to city council, petition of Mr. and Mrs. Stephen E. Baldwin for preservation of Pike Place Market as it is today, April 8, 1968, comptroller file no. 260566, SMA.

98. Melba Windoffer to city council, ibid.

99. Marilyn C. Skeels to city council, ibid.

100. K. Joey Grunden to city council, ibid.

101. Timothy Pettit to city council, petition of Timothy Pettit for preservation of the Pike Place Public Market in its present form, September 8, 1969, comptroller file no. 264592, SMA.

102. Robert M. Cour, "The Plight of the Pike Place Market, in Northwest Today," *Seattle Post-Intelligencer*, September 26, 1971.

103. "Civic Schizophrenia and the Friends of the Market," 1969, Victor Steinbrueck Papers, 1931–86, box 3, fol. 4, UWL-SC.

104. Ernst Gayden to city council, "Statement on the Pike Place Renewal Proposal," May 1, 1969, petition of Timothy Pettit for preservation of the Pike Place Public Market in its present form, September 8, 1969, comptroller file no. 264592, SMA.

105. Steven V. Roberts, "Seattle: A City with Growing Pains That Keeps Small Town Ways," *New York Times*, August 11, 1969.

106. *Seattle Business*, September 3, 1968, 4–5, UWL-SC.

107. Ibid., January 22, 1970, UWL-SC.

108. Ibid., July 26, 1971, 5–32, UWL-SC.

109. Braman to Steinbrueck, 1969, Victor Steinbrueck Papers, 1931–86, box 3, fol. 4, UWL-SC.

110. "News," Central Association of Seattle, August 12, 1970, UWL-SC.

111. Ibid.

112. Statement from Friends of the Market, November 25, 1970, press release, Victor Steinbrueck Papers, 1931–86, box 3, fol. 4, UWL-SC.

113. *Seattle Business*, January 25, 1971, UWL-SC.

114. David Brewster, "Seattle's Other Government," *Seattle Magazine*, December 1970.

115. "CHECC Challenged by Market Group," *Seattle Post-Intelligencer*, September 25, 1971.

116. Ada Louise Huxtable, "Keeping the There There," *New York Times*, May 16, 1971.

117. Ibid.

118. The *New York Times* seemed especially supportive of the Friends of the Market position; for a loving description of the market, see Douglas E. Kneeland, "Urban Renewal Threatens Seattle Market," *New York Times*, July 28, 1971.

119. Donald Aspinall Allan, "Seattle's Pike Place Market," *Gourmet*, June 1971, 20, 30.

120. Michael J. Parks, "Pike Plaza Publicity . . . or Is It Propaganda?" *Seattle Times*, September 12, 1971; "Everybody 'Loves' the Market," *Seattle Times*, September 12, 1971; "Toward Pike Plaza Accord," *Seattle Times*, September 21, 1971; "Confusion Intended," *Washington Teamster*, September 17, 1971.

121. "Mayor's Market Plan Opposed," *Seattle Post-Intelligencer*, September 21, 1971.

122. Mayor's proclamation re: passage of Initiative Measure no. 1 (C.F. 270105) at the November 2, 1971, general election—re: Pike Place Markets, city clerk files, 271529, December 1971, SMA.

Chapter 2: Neighborhood

1. "Environmental Health Services," *Annual Report* (Seattle-King County Department of Public Health, 1969), 30, Washington State University Library (WSUL).

2. "Seattle Model City Environmental Health Project, Progress Report," 1970, Environmental Health Project records, Model Cities Program, box 1, Seattle Municipal Archives (SMA).

3. "Environmental Health Project, Progress Report, 1970."

4. The Seattle-King County Department of Public Health published a photograph of the billboard publicizing the antirat campaign in "Environmental Health Services," *Annual Report* (Seattle-King County Department of Public Health, 1970), 26, WSUL.

5. See Bernard J. Frieden and Marshall Kaplan, *The Politics of Neglect: Urban Aid from Model Cities to Revenue Sharing* (Cambridge, Mass.: MIT Press, 1975), 14–67; Gareth Davies, *From Opportunity to Entitlement: The Transformation and Decline of Great Society Liberalism* (Lawrence: University Press of Kansas, 1996), 137–38; Marshall Kaplan, Gans, and Kahn, Inc., *The Model Cities Program: The Planning Process in Atlanta, Seattle, and Dayton* (New York: Praeger, 1970), 7–9; Carl Abbott, *Portland: Planning, Politics, and Growth in a Twentieth-Century City* (Lincoln: University of Nebraska Press, 1983), 192–206.

6. See *Citizen Participation Handbook, Seattle Model City Program* (Seattle: Seattle Model City Program, 1972), WSUL.

7. Kaplan, Gans, and Kahn, *Model Cities Program*, 41.

8. Quintard Taylor, *The Forging of a Black Community: Seattle's Central District, from 1870 through the Civil Rights Era* (Seattle: University of Washington Press, 1994); John Findlay, *Magic Lands: Western Cityscapes and American Culture after 1940* (Berkeley: University of California Press, 1992); Carl Abbott, *The Metropolitan Frontier: Cities in the Modern American West* (Tucson: University of Arizona Press, 1993), 1–29.

9. Kaplan, Gans, and Kahn, *Model Cities Program*, 41.

10. "Five Year Comprehensive Plan," draft, Seattle Model Cities Program, City of Seattle, September 23, 1968, 1, WSUL.

11. See Peter T. Nesbett, *The Complete Jacob Lawrence* (Seattle: University of Washington Press, 2000). In Seattle's Central District activists fought long and hard to obtain an old school in their neighborhood for a cultural center. The Northwest African American Museum opened in 2008 and now displays Lawrence's work from this period; see their Web site naamnw.org.

12. Taylor, *Forging a Black Community*, 6.

13. For more on the unique Asian American and African American makeup of the Central District, see Quintard Taylor, "Blacks and Asians in a White City: Japanese Americans and African Americans in Seattle, 1890–1940," *Western Historical Quarterly* 22, no. 4 (1991): 401–29.

14. Michael Egan and Jeff Crane, *Natural Protest: Essays on the History of American Environmentalism* (New York: Routledge, 2009). Robert Gottlieb has done the most to define a more diverse trajectory for this movement in his *Forcing the Spring: The Transformation of the American Environmental Movement* (Washington D.C.: Island, 1993). For examples of work on environmental racism and inequality, but less about urban actors as environmental activists, see Andrew Hurley, *Environmental Inequalities: Class, Race, and Industrial Pollution in Gary, Indiana, 1945–1980* (Chapel Hill: University of North Carolina Press, 1995); Robert D. Bullard, *Dumping in Dixie: Race, Class, and Environmental Quality* (Boulder, Colo.: Westview, 1990). For an excellent local example of the history of environmental inequality and its links especially to postwar urban growth, see Ellen Stroud, "Troubled Waters in Ecotopia: Environmental Racism in Portland, Oregon," *Radical History Review* 74 (1999): 65–95.

15. Linda Nash brings these aspects of environmental history together in *Inescapable Ecologies: A History of Environment, Disease, and Knowledge* (Berkeley: University of California Press, 2006).

16. Robert Self links suburbanization to the rise of black radical politics in Oakland during the 1960s in his book *American Babylon: Race and the Struggle for Postwar Oakland* (Princeton, N.J.: Princeton University Press, 2003). See also Adam Rome, "'Give Earth a Chance': The Environmental Movement and the Sixties," *Journal of American History* 90, no. 2 (Sept. 2003): 525–54 (esp. 527); idem, *The Bulldozer in the Countryside: Suburban Sprawl and the Rise of American Environmentalism* (New York: Cambridge University Press, 2001). In general the 1960s remain understudied in terms of the potential linkages between urban political organizing, suburban and urban political changes, and the rise of the environmental movement.

17. Andrew Kirk's excellent history of the counterculture environmental movement and Stuart Brand's influence on it makes the argument that the counterculture green movement was in part a product of a certain ethos of independent, libertarian, and do-it-yourself western activists and intellectuals; see Kirk, *Counterculture Green: The Whole Earth Catalog and American Environmentalism* (Lawrence: University Press of Kansas,

2007). For more on the persistence of this idea of western individualism despite the reality of federal government influence in shaping the West, see Richard White, *It's Your Misfortune and None of My Own: A New History of the American West* (Norman: University of Oklahoma Press, 1993).

18. Lyndon B. Johnson, "Special Message to Congress on Improving the Nation's Cities," *New York Times*, January 27, 1966.

19. Self, *American Babylon*, 139.

20. This was certainly true in Richard Daley's Chicago, where the mayor blocked the flow of federal money into programs that might have empowered local communities and undermined his own power.

21. Johnson, "Special Message."

22. Frieden and Kaplan, *Politics of Neglect*, 43.

23. Johnson, "Special Message."

24. James O. Wilson, *Urban Renewal: The Record and the Controversy* (Cambridge, Mass.: MIT Press, 1966), 491–508; Alison Isenberg, *Downtown America* (Chicago: University of Chicago Press, 2004), 166–202; J. Clarence Davies, *Neighborhood Groups and Urban Renewal* (New York: Columbia University Press, 1966), 1–29, 147–90.

25. See Frieden and Kaplan, *Politics of Neglect*, appendix.

26. "Model Cities Hoax," *Afro American Journal*, November 21, 1968.

27. Ibid.

28. Editorial, *Afro American Journal*, November 21, 1968.

29. Ibid., June 27, 1968; for a comparison with the Model Cities program in Oakland at the same time, see Self, *American Babylon*, 242–46.

30. Kaplan, Gans, and Kahn, *Model Cities Program*, 41.

31. Charles V. Johnson, interview by Trevor Griffey and Brooke Clark, videotape recording, March 3, 2005, Seattle, Seattle Civil Rights and Labor History Project, http://depts.washington.edu/civilr/johnson.htm.

32. Ibid.

33. Kaplan, Gans, and Kahn, *Model Cities Program*, 41.

34. "Model Area Planned," *Central Area Motivation Program Trumpet*, August 4, 1967.

35. "Central Area Task Forces: 'We Shall Overcome,'" special edition, *Trumpet*, October 12, 1967.

36. *Trumpet*, October 12, 1967.

37. "Hundley Will Direct Model-Cities Project," *Seattle Times*, December 19, 1967.

38. Interview with Walter Hundley by Mary Henrey, June 6, 1999, Seattle, Museum of History and Industry Library. For an example of the Model Cities logo—both image and the tag line "Make it Fly"—see any number of MC publications from the era, including the "Project Work Program," May 1972, Environmental Health, Seattle Model City Program, WSUL.

39. Kaplan, Gans, and Kahn, *Model Cities Program*, 45.

40. There are few detailed histories of Model Cities programs' effects. For a discussion of Model Cities in Albuquerque, see Amy Scott, "The Politics of Community in the Albuquerque Model Cities Program, 1967–1974," *New Mexico Historical Review* (forthcoming).

41. Self (*American Babylon*) compellingly argues that the African American community in Oakland, and especially the lucid critique of the Black Panther Party, was possible because this population saw the effects of the postwar period more clearly than did other neighborhoods and populations in the metropolitan area.

42. Don D. Wright, "Model City Program: Decent Home, Good Environment 'Impos-

sible Dream,'" *Seattle Times*, Sunday, April 6, 1969; Kaplan, Gans, and Kahn, *Model Cities Program*, 41–45.

43. Frieden and Kaplan, *Politics of Neglect*, 14–6 7.

44. Kaplan, Gans, and Kahn, *Model Cities Program*, 41–45.

45. Ibid.

46. Rachel Carson, *Silent Spring* (Boston: Houghton Mifflin, 1962); Jane Jacobs, *The Death and Life of Great American Cities* (New York: Random House, 1961); Ian L. McHarg, *Design with Nature* (Garden City, N.Y: Natural History Press, 1969); William S. Saunders, ed., *Richard Haag: Bloedel Reserve and Gas Work Park* (New York: Princeton Architectural Press, with the Harvard University School of Design, 1998).

47. Wright, "Model City Program."

48. Ibid.

49. See Seattle Model City Program (SMCP), *Evaluation Report: SMCP Programming in Housing and Physical Environment* (Seattle: Seattle Model City Program, 1971), WSUL.

50. Wright, "Model City Program."

51. Don. D. Wright, "Model City Program: Potential for Beauty Hurt by Pressure against It," *Seattle Times*, April 4, 1969.

52. Bower interviewed by Wright in Wright, "Model Cities Program: Potential"; Bob Santos, *Hum Bows, Not Hot Dogs: Memoirs of a Savvy Asian American Activist* (Seattle: International Examiner Press, 2002), 88–90, 91.

53. Wright, "Model Cities Program: Potential."

54. SMCP, *Evaluation Report*, 1.

55. Wright, "Model City Program: Potential."

56. SMCP, *Evaluation Report*, 6–11.

57. "Repair Service Project Work Program, 1971–72, Housing/Physical Environment," Environmental Health Project records, Model Cities Program, 1969–71, box 2, SMA.

58. Alice Staples, "Model Cities, the People's Program, and How It Works," *Seattle Times*, June 16, 1968.

59. "Introduction," *Annual Report* (Seattle-King County Department of Public Health, 1969), 2.

60. Ibid.

61. Don D. Wright, "Model City Program: Health Risks Climb Aboard Treadmill to Poverty," *Seattle Times*, April 1, 1969.

62. "Environmental Health Project C," December 13, 1968, 13, Environmental Health Project records, Model Cities Program, 1969–71, box 1, SMA.

63. "Addendum to Environmental Health Project C (Modifications to Grant Application)," May 19, 1969, 2, Environmental Health Project records, Seattle Model Cities Program, 1969–71, box 1, SMA.

64. Walter Hundley to Donald C. McDonald, M.D., acting regional program director, Community Health Services, May 19, 1969, Environmental Health Project records, Seattle Model Cities Program, 1969–71, box 1, SMA; "Environmental Health Project Expansion Proposal," August 17, 1970, Environmental Health Project records, Seattle Model Cities Program, 1969–71, box 1, SMA.

65. "Environmental Health Project Expansion Proposal."

66. "Environmental Health Project Summary," 1970, Environmental Health Project records, Seattle Model Cities Program, 1969–71, box 1, SMA.

67. Ibid.

68. "Seattle-King County Department of Public Health," Seattle Model City Environmental Health Project Progress Report, April 1, 1971–June 30, 1971, Environmental Health Project records, Seattle Model Cities Program, 1969–71, box 1, SMA.

69. "Seattle Model City Environmental Health Project Progress Report," 1970, 16.

70. "Environmental Health Project, Quarterly Report," 1970, Environmental Health Project records, Seattle Model Cities Program, 1969–71, box 1, SMA.

71. Charles V. Johnson, Interview.

72. "Environmental Health Project, Quarterly Report," 1970, 3.

73. "Health Advisory Board Monitoring Committee Visit to Environmental Control Project," April 8, 1970, Environmental Health Project records, Seattle Model Cities Program, 1969–71, box 1, SMA.

74. Ibid., 2.

75. Memorandum to Health Task Force from Monitoring Committee, Seattle Model City Program, September 4, 1970, Environmental Health Project records, Seattle Model Cities Program, 1969–71, box 1, SMA.

76. R. J. Webb, training and education officer, to Dr. Hugh Clark, Harborview Hospital, April 9, 1970, Environmental Health Project records, Seattle Model Cities Program, 1969–71, box 1, SMA.

77. "Environmental Specialties," 1970, Environmental Health Project records, Seattle Model Cities Program, 1969–71, box 1, SMA.

78. Pack Rats Association flyer, Environmental Health Project records, Seattle Model Cities Program, 1969–71, box 1, SMA.

79. Timothy Conlan, *From New Federalism to Devolution: Twenty-Five Years of Governmental Reform* (Washington, D.C.: Brookings Institution Press, 1998), 48.

80. Ibid; Wendell E. Pritchett, "Which Urban Crisis? Regionalism, Race, and Urban Policy, 1960–1974," *Journal of Urban History* 34 (Jan. 2008): 260–88.

81. "Model City Expansion Plans Are Revealed," *Seattle Times*, September 16, 1971.

82. "Model Cities Extension to Other Areas," *Seattle Post-Intelligencer*, September 21, 1971.

83. "Model Cities Leading a Healthy Life," *Seattle Post-Intelligencer*, October 3, 1971.

84. "Model Cities Expansion Voted," *The Medium*, October 21, 1971.

85. "Hundley Heads New Program," *Seattle Times*, October 30, 1971; interview with Walter Hundley. Hundley became the mayor's director of management and budget in 1974. In 1977 he became superintendant of the Seattle parks department; see Mary T. Henry, "Hundley, Walter R. (1929–2002)," at the HistoryLink Web site, www.historylink.org.

Chapter 3: Open Space

1. Parts of this scene were described by Lavern Hepfer, who participated in the "invasion" at Fort Lawton, in an interview with Edwin Samuelson, Lower Elwha Reservation, Washington, August 1970, American Indian Historical Research Project, University of New Mexico, Center for Southwest Research (hereafter AIHRPUNM); see also "Geronimo's Revenge," *Helix*, March 20, 1970, 4–5; "14 Indians Arraigned for 'Invasion,'" *Seattle Post-Intelligencer*, March 17, 1970.

2. At first they called themselves the "United Indian People's Council" but soon afterward adopted the name "United Indians of All Tribes."

3. "Geronimo's Revenge," 4–5.

4. Ibid.

5. Ibid.

6. The *Seattle Post-Intelligencer* reported, "The Seattle Liberation Front and the Black Panthers have offered their physical support and psychological clout to the Indians" ("How Indians Would Use Fort," March 22, 1970). The group also took their cause downtown, protesting at the federal courthouse and in full public view in the city; see "Indians Drum Up Support for Fort Claim," *Seattle Times*, March 10, 1970.

7. For an excellent presentation of the coalition's work to obtain the park, see Brian Gerard Casserly, "Securing the Sound: The Evolution of Civilian-Military Relations in the Puget Sound Area, 1891–1984," Ph.D. diss., University of Washington, 2007.

8. This context is not unique to Seattle. Hal Rothman's recent history of Golden Gate Park in San Francisco limns a similar debate over public space. Seattleites too argued about how best to use public open space, but their arguments exemplified how the meaning of nature and urban space were linked by the discourse of ecology. The conflict over the fort's future was one where local actors sought to wrest the space away from the government so they could shape it as they saw fit, revealing how the idea and political use of ecology evolved within an urban political context of the 1960s. See Hal K. Rothman, *The New Urban Park: Golden Gate National Recreation Area and Civic Environmentalism* (Lawrence: University Press of Kansas, 2004), 1–16, 17–30; see also Richard Walker, *The Country in the City: The Greening of the San Francisco Bay Area* (Seattle: University of Washington Press, 2007).

9. Adam Rome has argued that the "growing discontent of middle class women" and links between environmentalism and other political movements "contributed in key ways to the emergence of environmentalism" as a reinvigorated movement in the postwar era; see Rome, "'Give Earth a Chance': The Environmental Movement and the Sixties," *Journal of American History* 90 (2003): 525–54 (quotation on 527).

10. See Sale, *Seattle*; Ray T. Cowell, "Fort Lawton," *Washington Historical Quarterly* 19 (Jan. 1928): 31–36.

11. King County Commissioners to William C. Endicott, secretary of war, 1886, from *Fort Lawton: A Resource*, published by the U.S. Department of Interior, NPS Project Report, National Park Service, Northwest Region, in Pacific Northwest Forts and Blockhouses, Fort Lawton pamphlet file, University of Washington Libraries, Special Collections (UWL-SC).

12. Cowell, "Fort Lawton," 31–36.

13. Ibid; W. Clise, president of Seattle Chamber of Commerce, to the Board of Army Posts, Pacific Northwest Posts and Blockhouses, UWL-SC.

14. Sale, *Seattle,* 78–93.

15. Matthew Klingle captures important aspects of this emerging geography and the environmental history of Seattle in *Emerald City: An Environmental History of Seattle* (New Haven, Conn.: Yale University Press, 2007). For more on the City Beautiful movement generally and its influence in Seattle, see William H. Wilson, *The City Beautiful Movement* (Baltimore, Md.: Johns Hopkins University Press, 1989).

16. John C. Olmsted, "Special Report on the Improvement of Fort Lawton Military Reservation," 1910, 3–4, Seattle Board of Park Commissioners, Seattle, Washington.

17. Ibid.

18. This history is derived in part from Donald Voorhees's "Background Information on Fort Lawton," Seattle Audubon Society (hereinafter SAS), box 17, fol. 8, UWL-SC.

19. For a thorough recent history of metropolitanization in the United States, see Dolores Hayden, *Building Suburbia: Green Fields and Urban Growth, 1820–2000* (New York: Vintage, 2003), 3–18, 128–53.

20. "The White House Message on Natural Beauty to the Congress of the USA," February 8, 1965, published in *Trends in Parks and Recreation* 2, no. 2 (Apr. 1965): 1.

21. Ann Louise Strong, *Open Space for Urban America* (Washington, D.C.: Urban Renewal Administration, Department of Housing and Urban Development, 1965), iii.

22. See Rome, *Bulldozer.*

23. "Proposed New Site for 1961 World Fair Startles Officials," *Seattle Argus*, May 30, 1958.

24. "Will Fort Lawton Be Next Casualty of Economy Drive?" *Seattle Argus*, December 27, 1963.

25. Ibid.

26. See Kirkpatrick Sale, *Green Revolution: The American Environmental Movement, 1962–1992* (New York: Hill and Wang, 1993); see also Rome's detailed discussion of open space in *Bulldozer*, 119–52.

27. Stewart L. Udall, "Parks: The Challenge of Excellence," *Trends in Parks and Recreation* 1, no. 1 (July 1964): 1.

28. Richard A. Moore, ". . . On Beautification," *Trends in Parks and Recreation* 3, no. 3 (Oct. 1966): 7; *Trends* reflected the postscarcity thinking common to the era, as well as the ideas of the economist John Kenneth Galbraith and others who believed that the government's task in the postindustrial age was to manage work and leisure more rationally; see Galbraith, *The Affluent Society* (Boston: Houghton Mifflin, 1958), 1–5. For a discussion of the role experts played in the formation of early environmentalism, see Rome, *Bulldozer*, 1–13; Udall, "Parks," 1.

29. Strong, *Open Space for Urban America*.

30. "White House Message on Natural Beauty," 1.

31. Ibid.

32. Donald Goldman, "Conservation in the City," *Trends in Parks and Recreation* 3, no. 3 (July 1966): 25.

33. Moore, "On Beautification," 7.

34. Ibid., 8.

35. Ibid., 7–8.

36. Along with *Trends in Parks and Recreation*, other government agencies produced publications distributed to city planners, academicians, conservation groups, and urban decision makers during the mid-1960s. *Open Space for Urban America*, a publication produced by the Department of Housing and Urban Development in 1965, exemplified the coalescing conservation and open space movement. The publication, intended for the "informed layman and State and local officials," described the goal of the new conservation movement as "the creation of an urban society that retains a continuing association with the natural world." When open space advocates used the word *urban,* they meant everything from the old city core to the outward creep of metropolitan America. Open space advocates of this "new conservation movement" sought to preserve nature for recreation purposes and for "unspoiled" wilderness. But they spoke most clearly to those interested in creating "an orderly, efficient, and attractive urban environment within which men may enjoy the good life." The ideas found in such publications supported and encouraged urban leaders and the general public to begin to rethink not only their cities' land-use patterns but also the look and feel of the cities, if they had not begun to do so already. Publications such as *Open Space for Urban America* provided the rationale, described the planning process, gave examples of successful projects, and presented the federal role in open space preservation. Taking them together with the General Services Administration's handbook *Disposal of Surplus Real Property,* which described the process of reclaiming surplus federal land, Seattleites could begin to imagine the various alternative ways that they might creatively reuse Fort Lawton and other such places.

37. For more on the uneven distribution of wealth and resources built into the spatial patterns of the postwar metropolitanization, see Robert O. Self, *American Babylon: Race and the Struggle for Postwar Oakland* (Princeton, N.J.: Princeton University Press, 2003); Kevin Michael Kruse and Thomas J. Sugrue, *The New Suburban History* (Chicago: University of Chicago Press, 2006). In his discussion of the War on Poverty, Self de-

scribes the Great Society emphasis on "individual development" and "social disorganization" rather than jobs and desegregation. Beautification and the emphasis on the environmental problems in the city fit conveniently into this view of poverty as well; see Self, *American Babylon,* 1–20, 201.

38. See Steve Fraser and Gary Gerstle, eds., *The Rise and Fall of the New Deal Order, 1930–1980* (Princeton, N.J.: Princeton University Press, 1989). For another example of this struggle over political authority on city streets, see David Farber, *Chicago '68* (Chicago: University of Chicago Press, 1988), 169–77.

39. Women were also critical actors in pushing for Forward Thrust and were involved in the creation of METRO in the early 1950s.

40. For background on the early political organizing efforts of women in the Audubon Society and the links between consumption and environmentalism, see Jennifer Price, *Flight Maps: Adventures with Nature in Modern America* (New York: Basic Books, 1999), 1–109.

41. Stephen Bocking, *Ecologists and Environmental Politics: A History of Contemporary Ecology* (New Haven, Conn.: Yale University Press, 1997), 1.

42. Irene Urquart, "Field Trips," *Seattle Audubon Society Notes,* September 1964, UWL-SC.

43. See *Audubon Warblings,* April 1965, UWL-SC.

44. *Seattle Audubon Society Notes,* May 1965, 16, UWL-SC.

45. Ibid.

46. "Program Launched to Save Campus Marsh," *Seattle Audubon Society Notes,* November 1968, UWL-SC; see Audubon Programs for vols. 9–12, 1968–1972 UWL-SC.

47. Ibid.

48. Donna Hightower Langston, "American Indian Women's Activism in the 1960s and 1970s," *Hypatia* 18 (2003): 114–15.

49. After many years of court battles and civil disobedience, the fishing rights fight led to the famous Boldt decision, handed down on February 12, 1974, which affirmed Indian fishing rights. The Supreme Court reviewed and confirmed the lower court's decision in 1979. For more on the fishing rights controversy in Washington State in the earlier 1960s, see the volume produced by the American Friends Service Committee entitled *Uncommon Controversy: Fishing Rights of the Muckelshoot, Puyallup, and Nisqualy Indians* (Seattle: University of Washington Press, 1970), 107–46.

50. Susan Starbuck, *Hazel Wolf: Fighting the Establishment* (Seattle: University of Washington Press, 2002), 208–13.

51. Coll Thrush characterizes the Fort Lawton activists as a radical departure from preexisting Indian politics, but he does also describe how activists at Fort Lawton drew from local and national movements and support; see Coll-Peter Thrush, *Native Seattle: Histories from the Crossing-Over Place* (Seattle: University of Washington Press, 2007), 162–83. For instance, Bob Satiacum, a prominent voice in the 1964 fish-in controversy, became active in the Fort Lawton protest.

52. Thrush, *Native Seattle,* xiv–xv.

53. For more on Indian activism on the West Coast in the late 1960s and early 1970s, see Troy Johnson, Joanne Nagel, and Duane Champagne, eds., *American Indian Activism* (Urbana: University of Illinois Press, 1997); Paul Chaat Smith and Robert Allen Warrior, *Like a Hurricane: The Indian Movement from Alcatraz to Wounded Knee* (New York: New Press, 1996).

54. John Trudell and Earl Livermore, interview by D. Stanford and R. Lujan, Alcatraz, San Francisco, California, February 5, 1970, AIHRPUNM.

55. Ibid.

56. For more about these centers of activity, see Troy Johnson, *The Occupation of Alcatraz Island* (Urbana: University of Illinois Press, 1996); Susan Lobo, "Is Urban a Person or a Place? Characteristics of Urban Indian Country," *American Indian Research Journal* 22 (1998): 89–102. See also Coll Thrush's discussion of the women who organized the Seattle Indian Center, *Native Seattle*, 162–83.

57. Nicolas Rosenthal argues that this shift in political strategies was influenced not only by the radical movements of the larger youth culture but also by federal policy that increasingly supported "self-determination" and what he calls a politics of "Indianess" among urban Indians; see Rosenthal, "Repositioning Indianess: Native American Organizations in Portland, Oregon, 1959–1975," *Pacific Historical Review* 71 (2002): 415–38. Coll Thrush makes similar points about the shifting political winds in Seattle but also shows the continuity among these different groups; see Thrush, *Native Seattle*, 162–83. For a discussion of Whitebear's unique role as leader during this time, when the fish-in leader Bob Satiacum, for instance, attempted to lead the Fort Lawton without success, see Lawney L. Reyes, *Bernie Whitebear: An Urban Indian's Quest for Justice* (Tucson: University of Arizona Press, 2006), 86–107.

58. Thrush, *Native Seattle*, 162–83.

59. Herb Robinson, "Fort Lawton Plans May Be Changed," *Seattle Times*, March 15, 1967.

60. In November the organizations included the League of Women Voters, the Seattle chapter of the American Institute of Architects, Allied Arts, the Seattle Audubon Society, the Sierra Club, the Mountaineers, the Alpine Club, the Seattle chapter of the Federation of American Scientists, the Seattle Garden Club, the Izaak Walton League, and the Seattle King County Board of Realtors, among others. This partial list is taken from a November 18, 1968, letter to these groups from Donald Voorhees, SAS, box 17, fol. 7, UWL-SC.

61. "Ft. Lawton: Study in Civic Disillusion," *Seattle Post-Intelligencer*, October 11, 1968.

62. "The Battle of Fort Lawton," *Seattle Times*, October 13, 1968; Voorhees had considerable pull with Jackson and other senators; for the entire story of the CFLP's efforts, see Casserly, "Securing the Sound," 384–454.

63. *Seattle Audubon Notes*, September 1968, UWL-SC.

64. *Seattle Audubon Notes*, October 1968, UWL-SC.

65. Seattle Audubon Society, Conservation Committee, to Lieutenant General Alfred D. Starbird, August 29, 1968, SAS, box 17, fol. 7, UWL-SC.

66. Ibid.

67. "Marchers Go Forth to the Fort," *Magnolia Times*, November 28, 1968.

68. Henry M. Jackson, "Toward a National Land Use Policy," Seattle Planning Council, Seattle Washington, September 15, 1969, Henry M. Jackson Papers, box 234, fol. 5, UWL-SC.

69. "Jackson Sees Passage for Bill," *Seattle Post-Intelligencer*, May 30, 1969.

70. Ibid.

71. *An Act to Amend the Land and Water Fund Act of 1965*, Public Law 91-485, *U.S. Statutes at Large* (1970).

72. *National Environmental Policy Act of 1969*, Public Law 91-190, *U.S. Statutes at Large* (1970).

73. Joseph J. Shomon, director of the Nature Center Planning Division, to Mrs. Ann Mack, January 6, 1969, SAS, box 17, fol. 7, UWL-SC.

74. National Audubon Society, Nature Center Planning Division, *A Fort Lawton Park Nature Center: A Preliminary Proposal* (New York: Audubon Society, 1971).

75. "Indians Seized in Attempt to Take Over Coast Fort," *New York Times*, March 9,

1970; "Army Disrupts Indian Claim on Ft. Lawton," *Seattle Post-Intelligencer*, March 9, 1970; "Indians Drum Up Support"; "MP Now Knows How Custer Felt," *Seattle Post-Intelligencer*, April 3, 1970.

76. Hepfer, for instance, remembers a lot of people coming up from Alcatraz for the Seattle protest: "The people from Alcatraz, they were comin [*sic*] up, just about every plane that came brought people, and whenever they got some money, they would come up to help us" (Hepfer interview, AIHRPUNM).

77. According to the *Seattle Post-Intelligencer*, Donald Voorhees worked hard during this period to ensure the park's existence amid the publicity about Indian claims; see "Park on Fort Land Reaffirmed," *Seattle Post-Intelligencer*, April 3, 1970.

78. "How Indians Would Use Fort." Troy Johnson, a historian writing on the Alcatraz occupation, makes some attempt to place Native American protest within the context of other radical activism during the 1960s, but few other historians have made more explicit links between protest communities and their rhetoric; see Johnson, *Occupation of Alcatraz Island*. For work on Alcatraz, see Adam Fortunate Eagle's *Alcatraz! Alcatraz! The Occupation of Alcatraz of 1969–1971* (Berkeley, Calif.: Heyday Books, 1992); Johnson, Nagel, and Champagne, *American Indian Activism*; Smith and Warrior, *Like a Hurricane*; and, for a more popular treatment, Peter Matthiessan, *In the Spirit of Crazy Horse* (New York: Penguin Books, 1992). The term "ecological Indian" comes from Shepard Kretch's discussion of stereotypical ideas regarding the history of Native Americans' relationship to conservation in *The Ecological Indian: Myth and History* (New York: Norton, 1999). For an excellent discussion of the ways different groups of Americans have adopted Indian personas at different times, including the 1960s, see Philip Joseph Deloria, *Playing Indian* (New Haven, Conn.: Yale University Press, 1998).

79. Coll Thrush describes this activism as the creation of one "place-story" among many; see Thrush, *Native Seattle*, 180–81.

80. "How Indians Would Use Fort."

81. Ibid.

82. Philip J. Deloria points out this theme in *Playing Indian*, 154–80. Deloria includes a facsimile of the original People's Park manifesto, which was printed over a picture of Geronimo. Other People's Park leaflets also featured Native American imagery and references.

83. They used the concept of "right of discovery" (a phrase reminiscent of Western imperialism) in an ironic ploy to justify their claims, playing on the history of Indian-white relations to formulate their oppositional rhetoric.

84. "Park on Fort Land Reaffirmed," *Seattle Post-Intelligencer*, April 3, 1970.

85. Enclosed with letter from Voorhees to Henry M. Jackson, March 1969, Henry M. Jackson Papers, box 234, fol. 4, UWL-SC.

86. National Audubon Society, *Fort Lawton Park Nature Center*.

87. Dan Kiley, *Discovery Park, Seattle, Washington: Revisions to Fort Lawton Park Plan, November 1972* (Charlotte, Vt.: Dan Kiley and Partners, 1974).

88. Early in its existence the Daybreak Star center obtained funds from the state of Washington, as well as corporate and tribal sponsors—the Quinault tribe provided shakes for the roof and the Colville Confederated tribes provided logs—which were used to plan and construct the buildings. When the center opened in 1977, its funding came from both federal and local sources, particularly the Seattle Office of Economic Development and from the Seattle Arts Commission; see Reyes, *Bernie Whitebear*, 108–26.

89. "The Battle of Discovery Park," *Seattle Weekly*, July 27, 1983, 83.

90. Ibid., 25.

91. Ibid., 26.

Chapter 4: Ecotopia

1. For an excellent discussion of the Inland Empire and the imagined geography of eastern Washington and Spokane, see Katherine Morrisey, *Mental Territories: Mapping the Inland Empire* (Ithaca, N.Y.: Cornell University Press, 1997). For a discussion of world's fairs, see Robert W. Rydell, *All the World's a Fair: Visions of Empire at American International Expositions, 1876–1916* (Chicago: University of Chicago Press, 1984); idem, *World of Fairs: The Century-of-Progress Expositions* (Chicago: University of Chicago Press, 1993); John Findlay, *Magic Lands: Western Cityscapes and American Culture after 1940* (Berkeley: University of California Press, 1992), 214–64; R. W. Apple Jr., "Nixon Opens World's Fair in Spokane," *New York Times*, May 5, 1974; Phillip H. Dougherty, "Advertising: Spokane Expo '74," *New York Times*, March 29, 1974; Douglas E. Kneeland, "Expo '74 Gradually Replacing Spokane's Skid Row," *New York Times*, March 24, 1974. For more on Nixon's farm policy and the recent debates about agriculture, see Michael Pollan, *The Omnivore's Dilemma* (New York: Penguin, 2006), 51–52.

2. Robert Moses, New York's infamous planning czar, came to the fair, too, and approved of Seattle's efforts; see Lawrence E. Davies, "Seattle Is Hailed by Moses at Fair," *New York Times*, April 23, 1962.

3. J. Brooks Flippen, *Nixon and the Environment* (Albuquerque: University of New Mexico Press, 2000), 1–50.

4. James P. Sterba, "Expo '74: From Pleas to Save Nature to Sales Pitches for Vehicular Fun," *New York Times*, May 10, 1974.

5. "Epilogue: A Glance Back at Some Major Stories," *New York Times*, November 10, 1974.

6. Sterba, "Expo '74."

7. Apple, "Nixon Opens Fair"; for the history of New Federalism, see Timothy Conlan, *New Federalism: Intergovernmental Reform from Nixon to Reagan* (Washington, D.C.: Brookings Institution Press, 1988).

8. For details about the fair, including the sizable counterculture presence in the city and makeshift "people's park," see J. William T. Youngs, *The Fair and the Falls: Spokane's Expo 74: Transforming an American Environment* (Cheney: Eastern Washington University Press, 1996), 264–94.

9. Ibid.

10. Andrew Kirk, *Counterculture Green: The Whole Earth Catalog and American Environmentalism* (Lawrence: University Press of Kansas, 2008), 6.

11. Ibid., 8.

12. Mark Musick, "A Brief History of Tilth," typescript, 2002, in possession of the author; the event was inspired in part by Frances Moore Lappé's *Diet for a Small Planet* (New York: Ballantine Books, 1971), which placed food at the center of modern environmentalism and linked world hunger and injustice to this movement. Her work was influential in the Northwest and helped activists to frame a local response to problems then usually understood as national and global.

13. For a discussion of the concept of steady state and energetic systems theory, see Howard T. Odum, *Environment, Power, and Society* (New York: Wiley-Interscience, 1971); Howard T. Odum and Elisabeth C. Odum, *Energy Basis for Man and Nature* (New York: McGraw Hill, 1976).

14. Mark Musick, interview with author, September 19, 2008.

15. Ibid.

16. Ibid.

17. Wendell Berry to Gigi Coal and Bob Stigler, July 4, 1974, Tilth Association Papers (TAP), box 3, fol. 58, Washington State University, Manuscripts, Archives, and Special

Collections (WSU-MASC); when Berry visited the fair and met with the Northwest ecotopians, he was writing his influential book *The Unsettling of America: Culture and Agriculture* (San Francisco: Sierra Club Books, 1977).

18. Musick, "A Brief History."

19. Kirk, *Counterculture Green*, 13–42.

20. Howard T. Odum, *Environment, Power*; Odum and Odum, *Energy Basis*; the Odum brothers were the sons of Howard Odum, an important advocate of American regionalism during the 1920 and 1930s. Many of the sons' ideas about ecology, ecosystems, and region are linked to the earlier period of regionalism, a thread connecting the regionalist work of the 1970s with that of the 1930s. For more on Odum senior, see Robert L. Dorman, *Revolt of the Provinces: The Regionalists Movement in American, 1920–1945* (Chapel Hill: University of North Carolina Press, 1993), 29–142.

21. Odum, *Environment, Power*, vii. Indeed, many activists, such as Musick and Deryckx, worked within and with programs at Washington State University and Central Washington State University, and they helped to develop human ecology programs at Evergreen State College that addressed this issue in the early 1970s. In this way, the counterculture had the important assistance of mainstream government institutions. As the energy crisis of the 1970s heightened environmental fears about a coming apocalypse, Odum provided commonsense ideas for survival.

22. Odum, *Environment, Power*, 1.

23. Ibid., 7.

24. Ibid., 11.

25. Ibid.

26. Kirk, *Counterculture Green*, 143–44. One observer of these dispersed efforts, these terrarium-sized projects, compared such experiments to "lifeboats" or "arks" that would promote systems of survival in the event of environmental disaster; see My, *What Do We Use for Lifeboats When the Ship Goes Down?* (New York: Harper Colophon, 1976).

27. Although Tilth and similar counterculture organizations may have seen themselves as opposing mainstream or "big" science, their own interest in appropriate technology, as well as the support they received from universities, reveals how the cold war context—the boom in technology and high-tech industries in the West—had shaped the possibilities of new and local appropriate technologies by the mid-1970s.

28. New Alchemy, with West Coast and East Coast branches after 1970, would spread its message far and wide in popular 1960s magazines including *Ramparts*, the *Whole Earth Catalog*, and other publications gathering support for its dispersed projects in Massachusetts, California, Canada, and Latin America. New Alchemy eventually published the *Journal of the New Alchemists* and several books that directly inspired ecotopians in the Northwest. For this eager audience the group proposed that "centers of research and education be established" and that the knowledge gained raising fish or investigating "noncommercial varieties of nutritious foods" be "made available to all through community extension" (Nancy Jack Todd, ed., *The Book of the New Alchemists* [New York: E. P. Dutton, 1977], xiii–xviii).

29. Richard Merrill, Chuck Missar, James Buckley, and Thomas Gage, "Introduction," in *Energy Primer: Solar, Water, Wind, and Biofuels*, ed. Richard Merrill and Thomas Gage (Menlo Park, Calif.: Portola Institute, 1974), 2; the same year, RAIN, a Portland group that evolved from an organization called "Eco-Net," began publishing its newsletter, an organized version of the *WEC* containing regionally specific technology, garden, and community information for Portland activists and a national audience. RAIN also created a sourcebook—one of hundreds of such efforts throughout the country by the mid-1970s—

modeled on the *WEC*. Andrew Kirk details this burgeoning movement, pinpointing the influence of Stewart Brand's *Whole Earth Catalog* and subsequent Point Foundation as providing the critical infrastructure of ideas for these dispersed projects, especially in the West; see Kirk, *Counterculture Green*, 43–114.

30. "Ecotope Group," *Primer*, 190, TAP, box 3, WSU-MASC.

31. For more on this proliferation of counterculture environmentalism, see Andrew Kirk's work on the counterculture's embrace of alternative technologies and diverse tools for creating enduring environmental change in the "whole environment": Andrew Kirk, "Appropriating Technology: *The Whole Earth Catalog* and Counterculture Environmental Politics," *Environmental History* 6, no. 2 (Apr. 2001): 374–94; idem, "'Machines of Loving Grace': Alternative Technology, Environment, and the Counterculture," in *Imagine Nation: The American Counterculture of the 1960s and '70s*, ed. Peter Braunstein and Michael William Doyle (New York: Routledge, 2002), 353–78.

32. Mattew Klingle points out the darker side of the utopia, especially Callenbach's depiction of a racially divided Ecotopia, in *Emerald City: An Environmental History of Seattle* (New Haven, Conn.: Yale University Press, 2007), 231–32.

33. *Seriatim* 1 (Autumn 1977). For more on bioregionalism, see Dan Flores, "Place: An Argument for Bioregionalism," in *Northwest Lands, Northwest Peoples: Readings in Environmental History*, ed. Dale D. Goble and Paul W. Hirt (Seattle: University of Washington Press, 1999), 31–50.

34. Richard Merrill, *Seriatim* 1 (Autumn 1977).

35. Woody Deryckx, "Progress in Planning the Northwest Conference on Alternative Agriculture; Minutes from a Moveable Meeting," September 5, 1974, TAP, box 4, fol. 131, WSU-MASC.

36. Ibid.

37. William Cronon details the process of this alienation of city from nature in *Nature's Metropolis: Chicago and the Great West* (New York: W. W. Norton, 1991), 23–54.

38. Musick interview.

39. "Northwest Conference on Alternative Agriculture," November 21–23, 1974, promotional pamphlet, TAP, box 4, fol. 131, WSU-MASC.

40. Ibid.

41. "Directory of Folks at the Northwest Conference on Alternative Agriculture," November 21–23, 1974, TAP, box 4, fol. 121, WSU-MASC.

42. Todd, *Book of the New Alchemists*, vii. The New Alchemy Institute emerged in Massachusetts and later took hold on the West Coast. New Alchemy did not advocate the idea of ecotopia, but its ideas and especially the ideas of Richard Merrill encouraged Northwest groups that specifically sought to work in the area that they defined as a unique "bioregion."

43. Richard Merrill, "Krawlin' under a Dinosaur," *Seriatim* 1, no. 4 (Autumn 1977): 8.

44. Ibid., 9.

45. Ibid., 10.

46. "Directory of Folks."

47. Merrill's vision brought together "farm workers, the small farmers, the food co-ops, the backyard gardens" and the use of both "old" and "new" technologies when they were appropriate. The remainder of the conference merged social justice and environmental themes and resonated with the critical issues of the era: depleted fossil energy sources and urban deterioration (ibid., 10).

48. "Northwest Conference."

49. When Mark Musick sent out the final directory of participants in the months after the conference, he wrote: "A unique community of people gathered at Ellensburg,

Washington on November 21–23, 1974, for the Northwest Conference on Alternative Agriculture. More than 700 people attended, and we came from all over—from the cities, from the suburbs, from isolated farms and ranches, and from small towns scattered throughout the Northwest." The directory provided contact information for groups such as, Pacific Northwest Seed Bank/Network in Corvallis Oregon, Abundant Life Seeds in Seattle, and Energy Action of Washington in Seattle. The Northwest Trucking-Trade Network provided information about its alternative transportation network for produce and other goods, covering the region from Idaho and Washington to Oregon and California. "We came with a wide diversity of backgrounds, but with a common purpose: to explore the alternatives to corporate, chemical agriculture, and to seek new ways of working with each other and with the Earth" ("Directory of Folks at the Northwest Conference on Alternative Agriculture," TAP, box 4, fol. 121, WSU-MASC).

50. "Tools for Transition: Spring Quarter Program," Office of Continuing Education, Central Washington State University, 1976, TAP, box 3, fol. 82, WSU-MASC.

51. See Findlay, *Magic Lands*, 216–28.

52. Ibid.

53. The surrounding county grew by about 215,000 as Seattle lost population. The city's population would not reach that of the 1960s again until the mid-1990s, during the prosperity of the "dot-com" bubble; see "1970s Census" at http://www.historylink.org.

54. "Pike Place Market Farmer/Vendor Study," Pike Place Project, City of Seattle, Department of Community Development, May 1974, 3–9, UWL.

55. See Klingle, *Emerald City.*

56. YANIP (Yesler Atlantic Neighborhood Improvement Project) Report, October 1972, YANIP files, box 2, fol. 5, Seattle Municipal Archive (SMA).

57. Ibid.

58. She has since changed her name to "Darlyn Del Boca," but the name "Rundberg" appears in the public record from the period.

59. "A Look at Seattle's Back-to-the-Earth Plan," *Seattle Post-Intelligencer,* March 7, 1973. For more on the history of the P-patch and the city's community garden program, see http://www.seattle.gov/Neighborhoods/ppatch/.

60. See http://www.seattle.gov/Neighborhoods/ppatch/.

61. Chapman went on to become a founding member of the Discovery Institute, a think tank located in Seattle that has championed causes including intelligent design and a neo-ecotopian vision of an integrated northwestern regionalism called the Cascadia. See http://www.cascadiaproject.org/. ·

62. "A Look at Seattle's Back-to-Earth Plan," *Seattle Post-Intelligencer,* March 7, 1973.

63. For a perceptive contemporary discussion of cooperatives in the 1970s, see Daniel Zwerdling, "The Uncertain Revival of Food Cooperatives," in *Co-ops, Communes, and Collectives: Experiments in Social Change in the 1960s and 1970s,* ed. John Case and Rosemary C. R. Taylor (New York: Pantheon Books, 1979), 89–111; idem, "But Will It Sell?" *Environmental Action,* March 13, 1976, 3–9; see also Warren J. Belasco, *Appetite for Change: How the Counterculture Took on the Food Industry, 1966–1988* (New York: Pantheon, 1989), 69–108.

64. *Puget Consumers' Cooperative Bulletin,* August 1965, in the Puget Consumers' Cooperative archive, Seattle; because the systems for assigning volume and issue numbers changed, I use just the month and year in the following citations.

65. *Puget Consumers' Cooperative [PCC] Bulletin,* August 1965, February 1971.

66. Ibid., September 1974, October 1974.

67. Belasco, *Appetite,* 68–108.

68. "Major Obstacles Plowed under for City's P-Patch," *Seattle Post-Intelligencer*, April 11, 1973.

69. *PCC Bulletin*, September 1975; Binda Colebrook, *Winter Gardening in the Maritime Northwest* (Arlington, Wash.: Tilth Association, 1977).

70. *PCC Bulletin*, April 1971.

71. Ibid., October 1973.

72. "City to Sprout Green Thumbs?" *Seattle Post-Intelligencer*, March 6, 1973.

73. "An Ordinance Relating to the Department of Human Resources," Council Bill 94891, April 1, 1974, Engineering Department, unrecorded subject files, SMA.

74. *Tilth Newsletter* 1, no. 8 (1975): 17, TAP, box 4, fol. 122, WSU-MASC.

75. Ibid.

76. Memo, John Marsh to Gordin Lerimer, "Report on Community Gardens Northwest," February 11, 1975, Department of Parks and Recreation, superintendent files, box 46, fol. 16, SAM.

77. Edward S. Singles, director of Human Resources, to Councilman John R. Miller, January 17, 1974, "Pea Patch Gardens 1973–1974," Engineering Department, unrecorded subject files, SMA.

78. See "A Resolution Approving and Adopting for Inclusion in the City's 1974–1983 Capital Improvement Program the City's 'P-Patch' Project," Resolution 24682, September 16, 1974, SMA.

79. See Seattle City Council Ordinances 103172 and 102218, George Benson, subject files, box 18, fol. 6, SMA.

80. "Ripping up Concrete for Garden—a 'Natural,'" *Seattle Sun*, August 28, 1974.

81. "One P-patch Last Year, 15 Now," *Sunset Magazine*, June 1974, 222.

82. Carl Woestwind (formerly Woestendiek), interview with author, September 23, 2008, and October 7, 2008.

83. Woestwind interview, October 7, 2008.

84. Ibid.

85. Daniel Cohen, "Food, Learning Can Be the Fruits of the Good Shepherd Orchard," *Seattle Sun*, October 26, 1977.

86. Ibid.

87. Carl Woestendiek, "City Farming . . . Using the Nooks," *Tilth Newsletter*, Summer 1977, 5.

88. Ibid., 4–5.

89. Kate Louahened to John Vibber, project manager, March 1978, 5080-05, box 38, fol. 3, SMA.

90. "Goals and Objectives Statement," Seattle Tilth, circa 1977, TAP, box 1, fol. 13, WSU-MASC.

91. Woestwind interview, October 7, 2008.

92. C. M. Girtch to Walter Hundley, Department of Parks and Recreation, Facilities Maintenance Division, construction and maintenance files, box 38, file 3, SMA.

93. Carl Woestendiek, "Report from the Seattle Chapter," circa 1977, TAP, box 1, fol. 13, WSU-MASC.

94. "Introduction," TAP, box 1, fol. 13, WSU-MASC.

95. Carl Woestendiek to Tilth Membership, February 20, 1978, TAP, box 1, fol. 13, WSU-MASC.

96. "Ecotope Group," TAP, box 1, fol. 13, WSU-MASC.

97. Bob Santos, *Hum Bows, Not Hot Dogs! Memoirs of a Savvy Asian American Activist* (Seattle: International Examiner Press, 2002), 102; Bob Santos, interview with the author, November 17, 2008.

98. Quintard Taylor, "Blacks and Asians in a White City: Japanese Americans and African Americans in Seattle, 1890–1940," *Western Historical Quarterly* 22, no. 4 (1991): 401–29.

99. Santos interview.

100. Ibid.

101. Ibid.

102. See Pollan, *Omnivore's Dilemma*, 134–84. For examples of community gardens in other places, including New York City, see Delores Hayden, *The Power of Place: Urban Landscapes as Public History* (Cambridge, Mass.: MIT Press, 1997).

103. Santos, *Hum Bows, Not Hot Dogs*; Santos interview.

104. *Asian Family Affair*, the International District's activist newspaper, chronicled these events; see, e.g., *Asian Family Affair* 1, no. 7 (Oct. 1972), 1, no. 9 (Dec. 1972), 2, no. 1 (Feb. 1973); see also Santos, *Hum Bows, Not Hot Dogs*, 78–88. A hum bow is a savory meat-filled Chinese pastry.

105. Santos interview.

106. Santos, *Hum Bows, Not Hot Dogs*, 89.

107. Neil H. Twenkler and Associates to Darlyn Rundberg, "Inspection of Proposed Community Garden Plot," April 8, 1976, Engineering Department, unrecorded subject files, SMA.

108. Santos, *Hum Bows, Not Hot Dogs*, 103; Santos interview; Elaine Ko, interview with the author, December 31, 2008.

109. *Tilth Newsletter*, Winter 1977, 13, TAP, box 4, fol. 122, WSU-MASC.

110. Santos interview.

111. *Tilth Newsletter* 1, no. 2. (1975), TAP, box 1, fol. 13, WSU-MASC.

112. Ibid.

113. Ibid.

114. Santos interview.

115. Santos, *Hum Bows, Not Hot Dogs*, 106.

116. *Tilth Newsletter*, Winter 1977, 13, TAP, box 4, fol. 122, WSU-MASC.

117. Ibid; "Report on Open Space, Section 2: Agriculture Land as Open Space," King County Environmental Development Commission, March 1973, 5, UWL.

118. "Report on Open Space," 17–24.

119. "King County Agriculture Protection Program," 1–4.

120. "Pike Place Market Farmer/Vendor Study," 1; for more on this preservation moment in Seattle's urban past, see Judy Morley, *Historic Preservation and the Imagined West* (Lawrence: University Press of Kansas, 2006). *Seattle Magazine* gave voice to this progressive movement; see "We've Toyed with Seattle Long Enough," *Seattle Magazine*, February 1966, 15–19.

121. The environmental historian Theodore Steinberg uses this notion of the "organic city" to describe late nineteenth- and early twentieth-century cities before extensive sewer systems and garbage disposal and during an era of larger animal populations (horses, etc.) in the city; see Steinberg, *Down to Earth: Nature's Role in American History* (New York: Oxford University Press, 2002).

122. "Pike Place Market Farmer/Vendor Study," 8.

123. Morley emphasizes this curious arrangement where the city entered the farming business in order to ensure a crop at the market; see Morley, *Historic Preservation*, 125–26. See also "Pike Place Market Farmer/Vendor Study," 19.

124. "Pike Place Market Farmer/Vendor Study," 19.

125. "Welcome to Pragtree Farm," circa 1979, TAP, box 2, fol. 28, WSU-MASC.

126. Ibid.

127. *Tilth Newsletter*, Spring 1978, 2, TAP, box 1, fol. 122, WSU-MASC. Tilth's experience and efforts usually passed beneath the radar of public officials until 1980, when a King County Agricultural Task Force report on local agriculture, yet another dire warning about the decline of regional food-producing land, mentioned the ELT in its report on land use in the county and possible alternatives. The report described the ELT as "the non-governmental Evergreen Land Trust," which owned "farmland in Snohomish, Skagit and Whatcom counties." All the lands were "donated," and "because the land was financed several years ago, the land and interest payments [were] reasonable and the farmers were able to get into agriculture without a large down payment for land acquisition." The arrangement, according to the report, allowed the ELT to "hold its lands in perpetuity" but made it possible to "mortgage its properties in order to finance additional land acquisitions"; see "King County Agricultural Task Force Report on Local Agriculture," King County Office of Agriculture and the King County Cooperative Extension, December 1980, 75, UWL. Within the city, community gardens and Tilth's urban agriculture center spread the gospel of organic and small-scale farming, helping to create markets for the goods its members produced in the alternative network. On Pragtree farm, Tilth's Ecotopians experimented in the style of the New Alchemy Institute to develop alternative energy sources and regionally appropriate and hardy vegetables. In 1977 Tilth published its first research publication, *Winter Gardening in the Maritime Northwest*, based on research performed at the Pragtree farm with grants from members of Puget Consumers' Co-op in Seattle as part of "the Winter Garden Project." The result was a more dependable winter crop of organic vegetables for year-round markets in the region and the dissemination of know-how to local farmers and others "seeking to be as self-reliant as possible" (*Tilth Newsletter*, Spring 1978, 2, TAP, box 1, fol. 122, WSU-MASC).

128. *Tilth Newsletter*, Winter 1977, 12–13, TAP, box 1, fol. 122, WSU-MASC.

129. "Refugees Found Success—the Hard Way," *Seattle Times*, July 30, 1979.

130. Darrell Glover, "Laotian Refugee Farmers Receive a 'Thank You' and a Real Feast," *Seattle Post-Intelligencer*, February 21, 1983.

131. Ibid.

132. Marion Burros, "The Northwest's New-Found Riches: Its Own Fine Food," *New York Times*, April 17, 1985. The *Times* mentioned only Musick's efforts at Pragtree farm in the 1980s and nothing about Tilth's larger labors or hard-won network. Musick was featured as the innovative supplier of the "wild greens for the unusual salads served" at Seattle restaurants. The wild and cultivated greens raised on Pragtree farm included chickweed and wild chrysanthemum, upland cress blossom, wild sorrel, and arugula. By 1985 the wild greens project at the farm had begun to elicit interest among the foodies of Seattle and New York. The *Times* feature included recipes from Seattle's up-and-coming restaurants that used local produce. The article even included a recipe for "Pragtree Farm Salad Dressing."

133. Schuyler Ingle, "Goodbye Applets, Hello Arugula," *Seattle Weekly*, February 20–26, 1985.

134. Ibid.

135. Review in *RAIN Magazine*, August–September 1982, 11, TAP, box 6, fol. 194, WSU-MASC.

136. Morley, *Historic Preservation*, 91–126.

137. Mark Musick, draft of article for *In Business Magazine*, September–October 1991, TAP, box 4, fol. 100, WSU-MASC; the journalist and essayist Michael Pollan recently recounted Kahn's success (or sell-out) story in "Naturally," a thought provoking article about organic foods that appeared in the *New York Times Magazine*, May 13, 2001, 30.

138. Burros, "Northwest's New-Found Riches."

139. "A Metamorphosis at Pike Place Market?" *Seattle Times*, July 13, 2003.

Chapter 5: Home

1. "Welcome to the Fourth (and last) Urban Homestead Open House September 20–21, 1980," Jody Aliesan collection, box 1, University of Washington Library, Special Collections (UWL-SC).

2. Alf Collins, "Recycle, Conserve, Grow, Dry, Chop, Insulate . . ." *Seattle Times*, January 13, 1980.

3. See "'A State of Mind': Columns Urged Self Reliance," *Seattle Times*, January 27, 1981; in 1979 such ideas were still considered rather novel.

4. Almost all reply cards expressed positive attitudes about Aliesan's open houses. She received at least 800 postcards from the people who visited her series of open houses and many more letters in response to her weekend *Seattle Times* columns; see postcards, Jody Aliesan collection, box 1, UWL-SC.

5. For a good treatment of the history and culture of the 1970s, including discussion of the energy crisis and recession, see Bruce Schulman, *The Seventies: The Great Shift in American Culture, Society, and Politics* (New York: Free Press, 2001).

6. Karen Hannegan-Moore to Jody Aliesan, January 1980, "*Times*-responses," Jody Aliesan collection, box 1, UWL-SC.

7. The Urban Homesteading program, although using western and pioneering imagery, was a nationwide program focusing on inner cities in the eastern United States. In Washington State the program was active mostly in the outlying parts of King County and not in Seattle. In south King County the county government encouraged would-be homeowners to reclaim fifty-seven "HUD-owned vacant properties"; see "An Application to the United States Department of Housing and Urban Development for an Urban Homesteading Demonstration Program," submitted by King County, Washington, in partnership with the Housing Authority of the County of King (January 1977). Urban homesteading encouraged "sweat equity" and "urban conservation," replacing the older ethic of urban renewal; see James W. Hughes, *Urban Homesteading* (New Brunswick, N.J.: Rutgers University Press, 1975), 1–7.

8. See a discussion of similar efforts in Dolores Hayden, *Redesigning the American Dream: The Future of Housing, Work, and Family Life* (New York: W. W. Norton, 1985), 45–49.

9. Ernest Callenbach, *Ecotopia: The Notebooks and Reports of William Weston* (Berkeley: Banyan Tree Books, 1975), 1–16. In the 1990s Callenbach authored books about sustainable business, and he continues to advocate the ideas first described in his book; see his Web site, www.ernestcallenbach.com.

10. See Roger Sale, *Seattle Past to Present: An Interpretation of the History of the Foremost City in the Pacific Northwest* (Seattle: University of Washington Press, 1976), 173–252; Janet D. Ore, *The Seattle Bungalow: People and Houses, 1900–1940* (Seattle: University of Washington Press, 2007).

11. Jody Aliesan, interview with author, February 2003.

12. For more on the history of Seattle's radical past, see Sale, *Seattle*, 94–135; Dana Frank, *Purchasing Power: Consumer Organizing, Gender, and the Seattle Labor Movement, 1919–1929* (Cambridge: Cambridge University Press, 1994); Albert A. Acena, "The Washington Commonwealth Federation: Reform Politics and the Popular Front," Ph.D. diss., University of Washington, 1975.

13. Aliesan interview.

14. Ibid.

15. For more on gender, environmental history, and environmental activism, see Virginia Scharff, ed., *Seeing Nature through Gender* (Lawrence: University Press of Kansas, 2003), xiii–xxii, 3–19. For women and the counterculture see, Gretchen Lemke-Santangelo, *Daughters of Aquarius: Women of the Sixties Counterculture* (Lawrence: University Press of Kansas, 2009).

16. For a discussion of back-to-the-land communal experiments, see Timothy Miller, "The Sixties Era Communes," in *Imagine Nation: The American Counterculture of the 1960s and '70s,* ed. Peter Braunstein and Michael William Doyle (New York: Routledge, 2002), 327–51; Peter Coyote, *Sleeping Where I Fall: A Chronicle* (Washington, D.C.: Counterpoint, 1998).

17. For an excellent discussion of lesbian counterculture groups and nature in the 1970s, see Catherine Kleiner, "Nature's Lovers: The Erotics of Lesbian Land Communities in Oregon, 1974–1984," in *Seeing Nature through Gender,* ed. Virginia Scharff (Lawrence: University Press of Kansas, 2003), 242–62. Aliesan and other counterculture ecotopians hoped to link landscapes, to trace connections between urban and hinterland spaces; see William Cronon, "Kennecott Journey: The Paths out of Town," in *Under an Open Sky: Rethinking American Western Past,* ed. William Cronon, George Miles, and Jay Gitlin (New York: W. W. Norton, 1992), 28–51; idem, "The Nationalization of Nature," *The Nation and Beyond,* special issue, *Journal of American History* 86, no. 3 (Dec. 1999): 976; and a similarly inspired work about Seattle, Matt Klingle, *Emerald City: An Environmental History of Seattle* (New Haven, Conn.: Yale, 2007). Although not engaging the environmental movement in any detail in her work, Lizabeth Cohen provides a definition of "citizen consumer" useful to understanding how Seattleites began to see their consumer choices and private decisions as part of civic life; see Cohen, *Consumer's Republic: The Politics of Mass Consumption in Postwar America* (New York: Alfred A. Knopf, 2003), 1–18. See also Douglas C. Sackman, "Putting Gender on the Table: Food and the Family Life of Nature," in *Seeing Nature through Gender,* 169–93, for a discussion elaborating links between nature, gender, and everyday domestic spaces.

18. Andrew Kirk, *Counterculture Green: The Whole Earth Catalog and American Environmentalism* (Lawrence: University Press of Kansas, 2008), 1–12.

19. "Some Ideas on the 'Lifehouse' Concept," *Whole Earth Catalog* (Mar. 1970), 34. For a discussion of the shifting uses and meanings of the word *survival,* from the pre-1970s usage to the environmental-era usage, see Warren J. Belasco, *Appetite for Change: How the Counterculture Took on the Food Industry, 1966–1988* (New York: Pantheon, 1989), 30–31.

20. These words came from a leaflet advertising a teach-in about the park at which luminaries such as Jane Jacobs and Paul Goodman put their support behind a lineup of speakers that included members of Ecology Action, Todd Gitlin, the Radical Student Union, and city planners; see "The Politics of Ecology: A Teach-In To Support People's Park," Sixties Ephemera Collection, Bancroft Library, University of California, Berkeley.

21. Garrett De Bell, *The Environmental Handbook* (New York: Ballantine, 1970), 239. Jane Jacobs urged Americans to take another look at "slums" and existing housing structures of the city to find more human-scaled ways for curing the city's problems; see Jacobs, *The Death and Life of Great American Cities* (New York: Random House, 1961).

22. De Bell, *Environmental Handbook,* 240.

23. Ibid., 247.

24. Aliesan interview, February 2003; see http//www.nas.com/riverfarm/elthist .htm.

25. "Ecotope Group," circa 1977, Department of Parks and Recreation, ser. 38, fol. 3, Seattle Municipal Archives (SMA).

26. Ibid.

27. Ibid.

28. Tim Magee, "A Solar Greenhouse for the Northwest," and "Ecotope Group's Demonstration Solar Greenhouse," Department of Parks and Recreation, ser. 38, fol. 3, SMA.

29. "Ecotope Group."

30. For a discussion of the increasingly apocalyptic mood of environmentalists and a receptive, if overwhelmed, public during the 1970s, see Kirkpatrick Sale, *The Green Revolution: The American Environmental Movement, 1962–1992* (New York: Hill and Wang, 1993), 29–45; Schulman, *The Seventies.*

31. For more on this history of suburbs and the relationship of urbanism and nature, see Kenneth T. Jackson, *Crabgrass Frontier: The Suburbanization of the United States* (New York: Oxford University Press, 1985); Robert Fishman, *Bourgeois Utopia: The Rise and Fall of Suburbia* (New York: Basic Books, 1987).

32. Janet Ore, *The Seattle Bungalow: People and Houses, 1900–1940* (Seattle: University of Washington Press, 2007), 20–129; Klingle, *Emerald City,* 163–64.

33. Ore, *Seattle Bungalow,* 27.

34. *Citizen Participation Handbook: Seattle Model City Program* (Seattle: Executive Department, City of Seattle, 1972), 1, 29.

35. See *Fremont Forum,* March 1977, 1.

36. For a brief history of the "neighborhood revolts" in Portland, see Abbott, *Metropolitan Frontier.*

37. Carolyn Kelly, "Future," *Fremont Forum,* June 1977, 1.

38. For a local discussion of this experience of shifting from Model Cities to block grant funding, see "The Housing and Community Development Act of 1974: An Open Letter on the Future of Seattle," a supplement produced by the city of Seattle and published in the *Seattle Sun,* October 30, 1974.

39. Richard Nixon, "Radio Address about the State of the Union Message on Community Development," March 4, 1973, available at www.presidency.ucsb.edu.

40. "The Housing and Community Development Act of 1974," *Seattle Sun,* October 30, 1974.

41. Ibid., 2.

42. This same concern would resurface during the Commons debate in the 1990s.

43. "Seattle in Transition: Neighborhoods Come Alive," *Seattle Post-Intelligencer,* February 8, 1976.

44. *Fremont Forum,* March 1977.

45. Ibid.

46. For examples of these developments, see Klingle, *Emerald City,* 203–29.

47. Ibid., 3.

48. Eli Rashkov, "Downtown vs. Neighborhoods," *Fremont Forum,* December 1977, 5.

49. "A Sense of Community: Seven Suggestions for Making Your Home and Neighborhood a Better Place to Live," Seattle, Washington, Department of Community Development, Office of Neighborhood Planning, 1975, University of Washington Library (UWL); the federal government, through the federal extension service, also published pamphlets that encouraged both community-building organizations and ideas about "environmental quality and conservation" in neighborhoods; see Federal Extension Services, U.S. Department of Agriculture, *Community Resource Development: How Cooperative Development Helps* (Washington, D.C.: Government Printing Office, 1969); and Forest Service, U.S. Department of Agriculture, *People, Cities, and Trees* (Washington, D.C.: Government Printing Office, 1970; rev. 1974).

50. Hughes, *Urban Homesteading,* 1–7.

51. *Fremont Forum*, March 1977, 1.

52. Ibid.

53. "Feature," *Fremont Forum*, August 1977, 2; "Your Home: Before and After," *Fremont Forum*, October 1977, 3; "Rehab Co-op," *Fremont Forum*, December 1977, 3.

54. "Flash," *Freemont Forum*, March 1977, 4.

55. These young professionals were often called "yuppies"; for an excellent definition of the term, see Schulman, *The Seventies*, 241–46.

56. *Fremont Forum*, October 1977, 2–3.

57. After 1975 the housing values moved out of their five-year slump, and older homes "(those built before 1935) continued to appreciate faster than middle-aged or more recently constructed homes" (*Real Estate Research Report: For the City of Seattle and Metropolitan Area* 29, no. 2 [Autumn 1978]: 16, 17). These figures suggest the renewed interest in the middle city and the efforts of rehabilitation of older homes in Seattle.

58. *Fremont Forum*, October 1977, 3.

59. Ibid.

60. "Proposal to the City of Seattle: Research Project—Solid Waste Source Separation. Request for Grant Assistance, March 31, 1976, Submitted by: Fremont Recycling #1, Armen Stepanian and Dick Nelson," box 9, fol. 6, Legislative Department, SMA.

61. *Household Recycling: What to Recycle/How to Prepare It* (Seattle: Seattle Recycling Project, 1977), UWL.

62. For a discussion of recycling as an expression of consumer activism earlier in the twentieth century, see Cohen, *Consumer's Republic*.

63. "Recycling Is the Mayor's Bag," *Fremont Forum*, May 1978, supplement.

64. Ibid.

65. Bob Swanson, Committee on the Environmental Crisis, to Mayor Uhlman, recycling 1973, box 78, mayor's general files, UWL-SC; Norman C. Jacox, Electric League of the Pacific Northwest, letter and pamphlet ("Outa Sight") sent to Mayor Uhlman, recycling 1972, box 68, mayor's general files, UWL-SC.

66. Steven Bender to Mayor Uhlman, recycling 1972, box 68, mayor's general files, UWL-SC. In 1973, Bob Swanson, the Sierra Club's recycling committee head and chairperson of the Committee on the Environmental Crisis for the club's student section at the university, explained his group's concerns about recycling and the reuse of "a resource as many times as possible before it is returned to the environment where biological processes will break it down." Swanson urged the mayor to consider "natural resources" and the "limited amount of virgin wood fibers" for paper products, as well as the energy used to manufacture more (the letter was written on recycled paper). The mayor replied to Swanson that "the current energy crisis has probably convinced even the most unaware that we cannot go on forever in our present superconsumer economy. Fuels and material shortages will become the order of the day, and recycling will become imperative." He lauded the Sierra Club's efforts and explained that King County was at that moment involved in a "massive study" of the county's solid waste system in search of "alternatives," including "recycling and reuse" (Mayor Uhlman to Swanson, November 20, 1973, recycling 1973, box 78, mayor's general files, UWL-SC).

67. "Summary Interim Report: Solid Waste Management Study," November 1973, recycling 1973, box 78, mayor's general files, UWL-SC.

68. Ibid.

69. "Seattle Recycling Project Final Report: A Pilot Study of Voluntary Source Separation Home Collection Recycling, Performed for the City of Seattle," Department of Human Resources by Fremont Public Association, S-1, recycling 1973, box 78, mayor's general files, UWL-SC.

70. "Recycling Turnout Is High," *Lake City Journal,* December 1, 1976.

71. "Mining for the Urban Ore," *Lake City Journal,* October 6, 1976.

72. Ibid.

73. "From the Mayor," *Fremont Forum,* June 1977, 2.

74. Joan Firey to Armen Stepanian, April 25, 1977, "Seattle Recycling Project Final Report," AC-2; Helen and Gunnar Pemte to Seattle Recycling Project, undated, ibid., AC-8; Victoria Stevens to John Miller, June 23, 1977, ibid., AC-3.

75. For a discussion of the links between consumption and the shaping of the urban form, see Lizabeth Cohen, "Is There an Urban History of Consumption?" *Journal of Urban History* 29 (Jan. 2003): 87–106; Cohen makes interesting connections between suburbanization in the postwar period and the deterioration and change—both economically and physically—in the inner city.

76. Aliesan interview; Lizabeth Cohen describes this sort of activism in America's past as the expression of "citizen consumers," a way of organizing and asserting power, for women in particular, in the public sphere through private and collective acts of consuming; see Cohen, *Consumer's Republic,* 31, 41, 109.

77. Aliesan interview.

78. Ibid.

79. Ibid.

80. See Hayden, *Redesigning,* 45–49.

81. *The Integral Urban House: Self Reliant Living in the City* (San Francisco: Farallones Institute/Sierra Club Books, 1979), ix; the Farallones group had been active since 1972. For a description of their efforts and funding by Stuart Brand's Point Foundation, see Kirk, *Counterculture Green,* 115–55.

82. *Integral House,* viii.

83. Ibid., 10.

84. Ibid., 436 and frontispiece.

85. Aliesan interview.

86. Postcards, Jody Aliesan collection, box 1, UWL-SC.

87. Aliesan interview; "Welcome to the Fourth."

88. "Welcome to the Fourth."

89. "Relevant Household Expenses," Jody Aliesan collection, box 1, UWL-SC.

90. Postcards, Jody Aliesan collection, box 1, UWL-SC.

91. Ibid.

92. "Welcome to the Fourth."

93. Ibid.

94. Ed Hume, "Now's Time to Get Rid of Moss on Roof," *Seattle Times,* January 27, 1980.

95. Aliesan interview.

96. Ibid.

97. "Dear 'Urban Homemaker,'" Sally to Jody, June 28, 1981, Jody Aliesan collection, box 1, UWL-SC; Aliesan interview.

98. Nancy Herman to Jody Aliesan, January 21, 1980, *Times*-responses, Jody Aliesan collection, box 1, UWL-SC.

99. "An Up Hill Daytrip," *Equal Marriage* column, *Seattle Times,* April 6, 1980.

100. Jim Sanderson, "Macho Is Murder," *Seattle Times,* April 6, 1980.

101. Michael James Huddleston, letter to the *Seattle Times, Times*-responses, Jody Aliesan collection, box 1, UWL-SC.

102. Postcards, Jody Aliesan collection, box 1, UWL-SC.

103. "The Heat Pump Appears to Do the Impossible," *Sunset Magazine*, November 1977, 126–28.

104. For a discussion of *Sunset Magazine*, idealized images of the western home, and the construction of regional identity in an earlier period, see Elizabeth Carney, "Suburbanizing Nature and Naturalizing Suburbia: Outdoor-Living Culture and Land-scapes of Growth," *Western Historical Quarterly* 38, no. 4 (2007): 477.

105. "The Changing Western Home," *Sunset Magazine*, November 1977, 108.

106. "In the Paving Cracks, Alyssum and Much Else," *Sunset Magazine*, October 1977, 256.

107. "Gray Water Put to Work in Your Garden?" *Sunset Magazine*, May 1977, 268–69; "A Revolution in Greenhouses: Going All the Way Solar, Seven Examples," *Sunset Magazine*, March 1977, 96–99; "Landscape with Old Railroad Ties (or Even New Ones)," *Sunset Magazine*, February 1977, 66–69; "Owner Built with Partially Scrounged Wood," *Sunset Magazine*, September 1977, 116.

108. "Floor Plan Flip-flop," *Sunset Magazine*, October 1978, 162–63.

109. "This Small City Garden Was Only Recently Just a Concrete Driveway," *Sunset Magazine*, July 1974, 56–57.

Epilogue: Commons

1. John Hinterberger, "Park Here—Whispering Firs and Salmon Runs: A Different Sort of Downtown Space," *Seattle Times*, April 17, 1991. For a thorough overview of the commons chronology, see Lynne Iglitzin, *The Seattle Commons: A Case Study in the Evolution of an Urban Village* (Seattle: Institute for Policy and Management, 1993).

2. Iglitzin, *Seattle Commons*, 1–10.

3. John Hinterberger, "What Should We Name this Field of Dreams?" *Seattle Times*, May 1, 1991; idem, "Promenade Park?—Readers Respond to Hinterberger's Call for a Fitting Park Name," *Seattle Times*, May 15, 1991.

4. Hinterberger, "Promenade Park."

5. "Seattle Commons a Grand Vision," *Seattle Post-Intelligencer*, February 9, 1992.

6. Ibid.

7. Sean Wilentz, *The Age of Reagan: A History, 1974–2008* (New York: Harper, 2008); Mark Sidran, the city attorney at the time, became the focus of intense anger over his "civility laws," which seemed to target the homeless; see "Down in the Alley," *Seattle Weekly*, October 21, 1999. The historian Mike Davis captured the era's unease about urban environments during the late 1980s and early 1990s in *City of Quartz: Excavating the Future in Los Angeles* (New York: Vintage Books, 1992), 223, 226.

8. Population figures in Seattle reflected the rapid changes. The 1980s saw Seattle's population return to its 1960 height of half a million people and more—an increase of over 22,000 people in the 1980s alone. While Seattle grew in population, the adjacent suburbs in King County grew even more rapidly, by 235,000 during the 1980s. For a discussion of these census data, see "1980 Census: Population of Seattle Drops below 500,000 and That of King County Increases to 1,250,000 in 1980," at www.historylink.org.

9. See Richard Florida, *The Rise of the Creative Class* (New York: Basic Books, 2002); the rise of this "creative class" and its increasingly refined tastes were celebrated and advertised by the *Seattle Weekly* and its successful series of local guidebooks boosting Seattle-area and regional restaurants and businesses beginning in the 1980s; see *Seattle Best Places* (Seattle, Wash.: Sasquatch Books, 1983); David Brewster, *Northwest Best Places: Restaurants, Lodgings, and Tourism in Washington, Oregon, and British Columbia* (Seattle, Wash.: Sasquatch Books, 1986).

10. Lawmakers argued that it was in the interest of the state, its "citizens, communities, local governments, and the private sector" to "cooperate and coordinate with one another in comprehensive land use planning"; for the complete plan, see the Washington State legislature's Web site, specifically, "Legislative Findings, sect. 36.70A.010, Washington Growth Management Plan, 1970, available at www.leg.wa.gov. Over the protests of developers, the mayor, and the city council, Seattle voters even approved a cap on the height of Seattle's ever-soaring corporate skyscrapers in 1989. Seattle's mayor Greg Nickels recently proposed changing the cap. For a brief history of it, see "Nickels Wants Taller Downtown," *Seattle Times*, January 7, 2005.

11. Hinterberger's "grassroots" plan caught fire, just as the city was in the midst of finalizing a new comprehensive plan for urban development to be considered in 1994. At the time, the mayor and city planners imagined development plans that responded to the recent Growth Management Plan and to the cap on urban growth in the city. In 1992, when the project took the name "Seattle Commons" and influential citizens created "the Commons Committee," everyone involved in the project hoped to include their vision for this 278-acre section of the city into whatever the comprehensive plan would be. Under the leadership of Mayor Norm Rice, the idea of "urban villages" began to form a central feature of Seattle's plans for downtown redevelopment.

12. Robert T. Nelson, "Seattle Commons," *Seattle Times*, February 23, 1992.

13. See Dolores Hayden, *Building Suburbia: Green Fields and Urban Growth, 1820–2000* (New York: Vintage, 2003), 201–29; Douglas Kelbaugh, *Common Place: Toward Neighborhood and Regional Design* (Seattle: University of Washington Press, 1997).

14. Sim Van der Ryn and Peter Calthorpe, *Sustainable Communities: A New Design Synthesis for Cities, Suburbs, and Towns* (San Francisco: Sierra Club Books, 1986); Peter Calthorpe, *The Next American Metropolis: Ecology, Community, and the American Dream* (New York: Princeton Architectural Press, 1993). For more on the Farallones Institute's role in the counterculture green movement, see Alexander Kirk, *Counterculture Green: The Whole Earth Catalog and American Environmentalism* (Lawrence: University Press of Kansas, 2008). New Urbanism's vision of a return to walking cities, appropriate scale, and use of vernacular styles was famously lampooned in the *Truman Show*, a late-1990s film.

15. Don Canty, "UW Brainstorm Teams See Green in Seattle's Future," *Seattle Post-Intelligencer*, April 9, 1992; Douglas Kelbaugh and David Maurer, eds., *Envisioning an Urban Village: The Seattle Commons Design Charrette* (Seattle: Dept. of Architecture, College of Architecture and Urban Planning, University of Washington, 1992).

16. Kelbaugh, *Envisioning*, 47.

17. For instance, Fred Bassetti, a founder of the civic group Allied Arts, had helped lead grassroots and citizen efforts to protect Pioneer Square from redevelopment in the late 1960s. Still others, including Paul Schell (also a member of Allied Arts and a developer), had been active in Seattle's unique place-making efforts during the 1970s as part of the Department of Community Development; see Iglitzen, *The Seattle Commons*, 15.

18. Iglitzen, *The Seattle Commons*, 15.

19. Ibid., 18. The 1990s were a time of public-private partnership, and the Commons supporters presented the perfect example of this supposed synergy of private interest with public good. The private sector certainly came through. Foundations, individuals, and philanthropic organizations such as the Trust for Public Land (a product of Stewart Brand's Point Foundation) gave land and cash grants for the project.

20. "Cascade Community: People's Neighborhood or Commercial Resource," *Northwest Passage*, August 20–September 10, 1973, 8–9.

21. Ibid.

22. "Cascade Cries 'Help' as More Homes Fall," *Seattle Sun*, September 25, 1974. In this article, the middle-city alternative newspaper featured the story of an embattled family of thirty-year neighborhood residents who were about to be evicted. "It appeared," the *Sun* wrote, "that a new chapter in the death of a neighborhood would be written by the end of the month" (ibid.); the *Sun* called itself the paper for "the Capitol Hill, University District, Eastlake, Montlake, and Cascade communities." According to the *Housing in Cascade* report, "Since 1970 the residential population of Cascade . . . decreased from approximately 1076 to around 850," "few families with children under 18" resided in the neighborhood, and "a third of the population" was "between 45 and 64 years old and over a quarter are 65 and older," a distribution "quite different from Seattle as a whole but characteristic of low-income downtown residential areas"; see Richardson Associates, Northwest American, Northwest Environmental Technology Laboratories, and Seattle, *Housing in Cascade: Cascade Neighborhood Study . . . Prepared for the Seattle Department of Community Development* (Seattle, 1977).

23. "Cascade Cries 'Help.'"

24. "10-Year Plan for the Cascade Neighborhood," in "The Cascade Neighborhood Sustainable Community Profile—1995," Cascade Neighborhood Council and UW Center for Sustainable Communities, 1995, 7–8, University of Washington Library, Special Collections (UWL-SC).

25. Saul Alinsky, *Rules for Radicals: A Practical Primer for Realistic Radicals* (New York: Random House, 1971); R. V. Murphy, "Mr. Speaker: Chopp on Poverty, Power and Olympia Politics," *Real Change*, February 2004, 19.

26. "Hard-working Chopp Has 'Eye for the Prize,'" *Seattle Post-Intelligencer*, January 20, 2003.

27. "Helping Hand Has Muscle—Lawmakers Activism Is Potent Force," *Seattle Times*, May 26, 1997.

28. "The Great Seattle Land Rush: Have You Considered a Mobile Home?" *Seattle Weekly*, April 7, 1976, 9–10; "Seattle in Transition: Neighborhoods Come Alive," *Seattle Post-Intelligencer*, February 8, 1976.

29. "Seattle in Transition."

30. Richardson Associates, *Housing in Cascade.*

31. "Reviews Mixed to Commons Deal—Detractors Don't Want Housing Lost," *Seattle Times*, July 27, 1993.

32. "Skeptics Raise Questions at Hearing on Commons," *Seattle Post-Intelligencer*, July 30, 1993.

33. "Does the Public Get the Picture? Arts Group Says Commons Proposal Needs More Study," *Seattle Post-Intelligencer*, April 27, 1993.

34. "Commons Plan Comes Under Fire—Elitists Threaten Cascade Area, Activists Say," *Seattle Times*, February 24, 1993.

35. Small business owners were the other main opponents to the Commons. And like the housing activists who had alighted in Cascade during the 1970s to defend housing rights and the "authenticity" of a mixed-use neighborhood, these actors opposed the Commons on the ground that it would disrupt the mixed-use ecology of the south Lake Union area. According to one businessperson in the area, "This town has a wealth of parks and to bulldoze an important commercial area for a park is simply out of the question" ("Skeptics Raise Questions at Hearing on Commons," *Seattle Post-Intelligencer*, July 30, 1993). Similarly, Walt Crowley argued that the neighborhood was an "incubator for business energy" ("At Ground Zero of the Commons, Emotions Run High," *Seattle Post-Intelligencer*, August 25, 1995). Supporters of the Cascade neighborhood had long nurtured their own idea of sustainability—be it business or housing.

36. "10-Year Plan," 7–8. The champions of Cascade emphasized a vision of planning on a small scale: an idea of the unique neighborhood of small businesses, humble and historic structures, and mixed uses, not the top-down vision of the supposedly elite planners. Matthew Fox, the head the Seattle Commons Opponents Committee (SCOPE), described the Common's vision as "top-down, tear-it-down urban renewal." "If the Commons supporters prevail," Fox explained, "this historic working-class neighborhood would be demolished to create a front lawn for luxury apartments and condominiums" ("Commons Foes, Backers Square off at Meeting," *Seattle Times*, July 27, 1995); see also "Commons Foes to Start Campaign," *Seattle Times*, July 25, 1995.

37. "10-Year Plan."

38. For an example of this work in other urban contexts, see Giovanna Di Chiro, "Steps to an Ecology of Justice: Women's Environmental Networks across the Santa Cruz Watershed," in *Seeing Nature through Gender*, ed. Virginia Scharff (Lawrence: University Press of Kansas, 2003), 221–41.

39. "10-Year Plan."

40. Ibid.

41. David Brewster, "Weekly Wash," *Seattle Weekly*, May 15, 1996.

42. Paid advertisement, *Seattle Weekly*, May 15, 1996.

43. "With Defeat, All Eyes on Allen," *Seattle Times*, May 23, 1996.

44. "Remaking South Lake Union," *Seattle Post-Intelligencer*, October 20, 2004.

45. "Remaking South Lake Union."

46. "Allen Raises Rents, Dander, Suspicions," *Seattle Times*, April 4, 1997.

47. For a description of his holdings, as well as failed and successful projects, see Mike Francis, "Paul Allen Slowly, Surely Steps into Public Light," *Oregonian*, August 14, 1994; Elizabeth Lesly, "Paul Allen: New Age Mogul," *Business Week*, November 18, 1996. For a more recent project, see J. Lynn Lunsford, "Rutan's Rocket Becomes First Private Manned Craft in Space," *Wall Street Journal*, June 22, 2004.

48. Linda Keene, "The Enigmatic Paul Allen," *Seattle Times*, June 11, 1997.

49. Allen remains elusive. Unlike other actors with the power to completely reshape the urban landscape—Bogue or Thomson, for instance—Allen is all but absent from the public stage and process. When Allen began to concentrate his energy and wealth on high-profile projects, he did so slowly and cautiously. Profiles from the mid-1990s all shared the assessment that, as one reporter put it, the billionaire was a "private man," a "shy" public figure who was "reclusive" and reluctant to become involved in public affairs. The reporter also remarked that Allen preferred the company of his "intensely close-knit family"—the stereotypically reserved and private Seattleite. See Keene, "The Enigmatic."

50. See the Alcyone Apartments Web site at http://www.alcyoneapartments.com/.

51. Ibid.

52. Ibid.

53. "Remaking South Lake Union."

54. George Howland Jr., "Vulcan Mind Meld," *Seattle Weekly*, August 27, 2003.

55. Ryan Chittum, "From Neglect to Biotech," *Wall Street Journal*, February 16, 2005.

56. The Alcyone community gardens are conspicuously called "Pea Patches," not "P-Patches," which is the designation of the citywide public system.

57. Douglas Sackman, "Putting Gender on the Table: Food and the Family Life of Nature," in *Seeing Nature through Gender*, ed. Virginia Scharff (Lawrence: University Press of Kansas, 2003), 187.

58. As William Cronon argues, the idea of wilderness that has so long undergirded environmental thought and activism tends to place nature outside history and thus the

city and people outside nature. In these formulations, he writes, "home" and the domestic garden appear as more constructive metaphors for establishing the codependency of humans and the natural world. "We need to discover a common middle ground in which all of these things, from the city to the wilderness, can somehow be encompassed in the word home," he maintains. "Home, after all, is the place where finally we make our living. It is the place for which we take responsibility, the place we try to sustain so we can pass on what is best in it (and in ourselves) to our children"; see William Cronon, "The Trouble with Wilderness; or Getting Back to the Wrong Nature," in *Uncommon Ground: Rethinking the Human Place in Nature,* ed. Cronon (New York: W. W. Norton, 1995). For a popular discussion of the garden as a metaphor, see Michael Pollan, *Second Nature: A Gardener's Education* (New York: Delta, 1991).

59. Cronon, "Trouble," 89.

60. For a description of the way neighborhood activists saw the distinction, see "10-Year Plan," 7–8.

61. Recent histories that engage the development of neighborhood politics and the segregation of public space—especially with regard to the racial politics of the postwar period—include Robert O. Self, *American Babylon: Race and the Struggle for Postwar Oakland* (Princeton, N.J.: Princeton University Press, 2003); Becky Nicolaides, *My Blue Heaven: Life and Politics in the Working-class Suburbs of Los Angeles, 1920–1965* (Chicago: University of Chicago Press, 2002). See Matthew Klingle's discussion of these episodes in Seattle in *Emerald City: An Environmental History of Seattle* (New Haven, Conn.: Yale University Press, 2007); for a discussion of this period within the national context, see Bernard Frieden and Lynne B. Sagalyn, *Downtown Inc.: How American Rebuilds Cities* (Cambridge, Mass.: MIT Press, 1989).

.

Index

Note: Page numbers in italic type indicate photographs or illustrations.